£1.75

KU-319-005

CONCISE CHEMICAL THERMODYNAMICS

Concise Chemical Thermodynamics

J. R. W. WARN

Senior Lecturer in Physical Chemistry and Instrumentation
Middlesex Polytechnic

Van Nostrand Reinhold (UK) Co. Ltd.

Published by Van Nostrand Reinhold (UK) Co. Ltd. Molly Millars Lane, Wokingham, Berkshire, England

First published 1969
Reprinted 1971, 1972, 1974, 1975
1976 paperback edition
Reprinted 1977, 1978, 1979, 1980, 1981, 1982, 1984

Library of Congress Catalog Card No. 73–83656
ISBN 0 442 30209 6

Printed in Great Britain by Henry Ling Ltd.,
at the Dorset Press, Dorchester, Dorset

Preface

Thermodynamics, like classical music, is an acquired taste. The initiation must be sensitively carried out, otherwise the mathematical rigour, like the formal structure of the music, acts to discourage a deeper relationship. It is sad but true that some students, both of thermodynamics and of Bach, never recover from the initial shock. In this, we are all losers. In this book, therefore, I have tried to present thermodynamics in a way which uses little mathematics, but which includes many practical and lively examples. I hope it will provide a basis for an introductory course at Honours Degree level, and will also suffice for Higher National Diploma and Certificate students.

In general, my aim has been to develop an understanding of Gibbs free energy fairly early on and then to apply this concept in several different fields. I have used many examples from chemical industry, in the firm belief that the basic *usefulness* of the subject must be demonstrated. I have therefore included a brief discussion of Ellingham diagrams in Chapter 10 and for a similar reason have omitted any mention of the Carnot cycle. I have used symbols consistent with the recommendations of the International Union of Pure and Applied Chemistry and have adopted units of the Systéme International d'Unités (SI). Accordingly, all thermochemical data are given in joules not calories. At the time of writing, general agreement has not been reached on the replacement of the atmosphere and gram mole by the bar and kilogram mole, and so the older units have been retained.

During the writing of this book, I have drawn on the experience and advice of many persons and organizations. I am very grateful to Drs John Golden and Ron Perrett for the reading of major parts of the manuscript, and for helpful comments and discussions. In the writing of Chapter 10, I was much indebted to personnel of Magnesium Elektron Limited, British Hydrocarbon Chemicals Limited, and the British Iron and Steel Research Association (who provided data for Fig. 10.1). Thanks are also due to the Dow Chemical Company (for permission to reproduce sections of the JANAF tables), to the National Bureau of Standards, Washington (for Fig. 3.1 and for data from which Table 2.1 was compiled) and to the Royal Institute of Chemistry (for permission to include questions from previous examination papers, which appear herein as Problems 7.5, 7.7, 8.5, 9.8, and 9.11). Miss Hazel Hawkes, who typed the manuscript against great odds, also has my gratitude. Notwithstanding all of this help, I would never have completed this task without the continuous and firm support of my wife and family; to them I am indeed very grateful.

My publishers have worked long and well in the removal of errors and ambiguities from the manuscript. Possibly, however, there may be some which remain, which I have hidden too well for detection by even their practised eyes. To me must fall the blame for these errors; I would however be glad to know of those detected by readers.

J. R. W. Warn

November 1968
Hertford

Symbols and Abbreviations

The symbols and abbreviations given below are based on the *Manual of Symbols and Terminology for Physicochemical Quantities and Units*, soon to be published by the International Union of Pure and Applied Chemistry.

a	activity
c	concentration
c	velocity of light
(c)	crystalline state
°C	degrees Celsius
C_P	molar heat capacity at constant pressure
C_V	molar heat capacity at constant volume
e	unit negative charge
E	electromotive force
F	Faraday's Constant
f	fugacity
g	gram
G	Gibbs free energy (F in some books by American authors)
(g)	gas phase
H	enthalpy (heat content)
i	electric current
k	Boltzmann constant
K	equilibrium constant
K	kelvin
l	litre
(l)	liquid phase
log	decadic logarithm
ln	exponential logarithm
m	molality
M	molar concentration
N	Avogadro's number
n	number of moles
P	pressure
p	partial pressure
Q	activity quotient
q	thermal energy added to system
R	universal gas constant
S	entropy
t	time

T	temperature
U	internal energy (E in some books by American authors)
V	volume
w	work done on system
x	mole fraction
α	coefficient of expansion
γ	activity coefficient
Δ	change of function of state, final less initial
μ	chemical potential
Σ	summation symbol

Contents

1

Energy

1.1 The Realm of Thermodynamics

Our world is characterized by a multitude of natural phenomena. It is a world of change, of movement and energy; of storms, earthquakes, cosmic rays, and solar flares. The range and complexity of these changes is so great that it would seem the height of foolishness to attempt to find a common thread, or a theme, that runs through all of them. None the less, over the centuries, by patient and careful observation, and aided by the occasional flash of insight, men have come to a partial understanding of the factors involved. At the heart of such a science is the concept of **energy.** Thermodynamics, as we now know it, derived its name originally from studies of the 'motive power of heat', and had to do primarily with steam engines and their efficient use. The main reason that the idea of energy was so poorly understood is that energy, the *capacity for doing work*, appears in so many different forms. However, after heat and work were understood to be but different forms of the same thing, and calculations of the efficiencies of steam engines were carried out, thermodynamics was applied more generally to all changes, both chemical and physical.

Thermodynamics is a science of the *macroscopic world.* That is, it requires no prior understanding of atomic and molecular structure, and all its measurements are made on materials *en masse.* This is not to say that an understanding of molecular phenomena cannot help us to grasp some difficult concepts. The branch of the subject known as statistical thermodynamics has assisted greatly in our understanding of entropy, for example, but the basic theories of thermodynamics are formulated quite independently of it. This point is even more evident if we consider a *complete description* of, say, the steam in a kettle of boiling water. A description in molecular terms would involve the position and nature of each particle, and its velocity at any instant. As there would be well over 10^{21} molecules present, this would be a humanly impossible task. On the macroscopic scale, however, we are glad to find that the chemical composition of steam, its temperature, and pressure, for example, are quite sufficient to specify the situation.

If we accept that such *variables* as temperature, pressure, and composition are a sufficient description of such a system, and if we are prepared to follow energy in all its various disguises, then we shall find that thermodynamics is a reliable pathfinder and guide to new and unexplored phenomena. We shall find that by taking relatively simple measurements such as heats of reaction and specific heats, we can predict the outcome, and even calculate the equilibrium constants, of changes which may never have been attempted before. Thermodynamics is a reliable guide in industrial chemistry, in plasma physics, in space technology, and in nuclear engineering, to name but a few applications.

We shall now discuss some aspects of the two main fields of application, which spring from the first and second laws of thermodynamics. The first law is the energy conservation law, which requires a clear understanding of energy's disguises. The second law deals with the concept of entropy, an increase of which may be regarded as one of nature's two fundamental driving forces. (The other driving force is the minimization of energy.)

1.1.1 Energy Bookkeeping

The rate of physical and mental development of the species *homo sapiens* was previously limited by naturally occurring genetic processes. Revolutionary extensions to man's faculties have been made in the last two centuries. They are the extension of his mental capacity which came with the introduction of the electronic computer in the mid-twentieth century, and the extension of his physique, which came with the development of machinery in the Industrial Revolution. We shall be concerned with this second aspect.

There is a close link between the per capita consumption of energy, and the state of physical advancement of a nation. Thus in 1700, the English population used only a thousand pounds of coal or its equivalent per head per year. In 1965 this figure had risen to five and a half tons, only a small fraction of which was needed for winter heating. The current consumption in the United States is over eight tons.

The world need of cheap and plentiful energy will continue to grow, and it is essential that we should understand how far the laws of thermodynamics can help to clarify the various processes of energy conversion, or assist us in making efficient use of energy.

Consider some examples of energy conversion, both present and future.

(a) Our muscular energy (as a *mechanical*, or work-doing form of energy) springs from the controlled combustion of carbohydrate foods. Some of the energy is used to warm us; some is used as mechanical energy. It also seems clear that a form of electrical energy is involved at an intermediate stage. Thus life itself depends on *energy conversion.*

(b) Plant life depends on converting radiant energy from the sun into the chemical energy of sugars and carbohydrates, which are photosynthesized from water and carbon dioxide.

(c) Our most basic natural commodity, fresh water, is scarce. Rivers are overloaded with pollutants, and water is often used and reprocessed several times between river source and sea. Cheap nuclear energy will soon be harnessed on a large scale to desalinate sea water. By using the recently developed multistage flash distillation process, a nuclear plant could produce 400 megawatts of electricity at 0.18 new pence per kilowatt hour, and simultaneously produce $(2.5)10^6$ cubic metres a day of pure water at a cost of only four pence per cubic metre. Such plants are in an advanced stage of preparation.

(d) Copper is present in the earth's crust to the extent of only seventy parts per million, and although it does occur naturally as boulder copper, all large deposits of this kind have long ago been exploited. It is now economically feasible to work low-grade ores containing under one per cent of copper. However, the layer of the earth immediately below the solid crust, the thiophil, probably contains substantial amounts of sulphides of copper and other metals. Although aluminium can be used in place of copper for many purposes, demand for copper will probably be maintained for many years. While known reserves are consumed, energy will become cheaper; thus, although deep mining is not at present necessary, it must be considered seriously in the future.

The key to the processes outlined above is the abundant supply of energy. In the years to come, this requirement will be met from a variety of sources, and many different processes of energy conversion will be used. One thing, however, is certain. At all stages in the planning and execution of these schemes, basic thermodynamic concepts will form a central part. Such is the main theme of this book.

1.1.2 Nature's Driving Forces

In the 1880s it was found that the combustion gases issuing from blast furnaces contained large amounts of carbon monoxide. The reactions in which monoxide is converted to dioxide, such as

$$Fe_2O_3 + CO \rightarrow 2FeO + CO_2$$

or

$$FeO + CO \rightarrow Fe + CO_2$$

were not therefore occurring to completion. This seemed to indicate serious inefficiency in the furnaces. Perhaps insufficient time was being allowed for the reaction to be completed, or possibly these reactions were basically reversible, with small equilibrium constants for the forward reactions. If the smelters had known more about thermodynamics, they would have realized that the second solution was correct, and many firms would have been spared the expense of increasing the height of their blast furnaces. They did this in the vain hope that further opportunity would be given to the reactants to achieve equilibrium; in fact the equilibrium could not be improved upon.

The *tendency* of a certain process to occur or not may be expressed in terms of the equilibrium constant. If we study the results of simple thermal measurements, thermodynamics will help us to calculate equilibrium constants of processes which have not been previously attempted, or which are difficult to study directly. It can answer the question '*how far*?' but is *quite unable to answer* '*how fast*?' A host of industrial processes, such as the cracking of light naphtha to form ethylene, or the synthesis of ethanol from ethylene and water vapour, depend initially on a very careful analysis of the equilibrium situation over a wide range of temperatures. Only when the feasibility of a new process has been demonstrated, by thermodynamic methods, is it necessary to build pilot plants, and choose catalysts. *Once a reaction strikes the rocks of an unfavourable equilibrium constant, no mere catalyst can salvage the wreck.*

1.2 Setting the Scene; Basic Ideas

Thermodynamics depends on a careful logical analysis of natural phenomena, and so we must first agree on a small number of basic terms.

1.2.1 System and Surroundings

The experiments we shall discuss will all be carried out in a small well-defined space, usually in the laboratory, which we can study closely; this small region of space is called the **system.** The system may be simple or complex; it may be, for example, a container of chemicals ready to react, or a simple vessel of gas. There may well be facilities for stirring (to bring about homogeneity), or for measuring temperature or pressure. Sometimes the boundaries are only imagined, but they can always be represented on paper, so that everything outside the system is differentiated as the **surroundings.**

A system may undergo alteration from outside in two distinct ways. From time to time, either *chemical matter*, or *energy* in one of its forms, may move across the boundary. The energy may be mechanical energy transmitted by a shaft, electrical energy flowing through wires, or conducted, or radiated, thermal energy. The system is said to be **closed** if no matter is allowed to move across the boundary, and **adiabatic** if no thermal energy passes into or out of the system. If neither matter nor energy is transferred, it is said to be **isolated.**

1.2.2 Functions of State

When describing an experimental sample of say two moles of methane, it is only necessary to specify a certain minimum number of variables in order to describe the system completely. Thus temperature and pressure would be quite adequate, although we could substitute volume, or density for one of them. In fact any two would automatically fix the remainder. These **functions of state,** or state variables as they are sometimes called, are closely dependent on one another, and describe the *state* or *condition* of the system as it is now, without reference to its immediate history. This 'lack of memory' has an important consequence. If for example we are measuring the change of temperature of a tank of liquid heptane, from a 'state 1', to a 'state 2', the value of the temperature difference, $T_2 - T_1$, is quite independent of the intermediate steps of the change.

If we were considering the amount of thermal energy which had been added to achieve this rise in temperature from T_1 to T_2, we would find that *it depended on the way in which it was accomplished.* In one case we might have raised the temperature by passing hot water through some heating coils. Alternatively, we might have used mechanical energy to agitate the liquid heptane; this too would result in a temperature rise, although no 'heat' or thermal energy would have been added. Thermal energy and mechanical energy themselves *would certainly not qualify* as functions of state, but could alter with the whim of the experimenter, and would depend on the *route* by which he chose to go from (state 1) to (state 2).

It should be pointed out here that a change of volume or of any other function of state is always measured as that of the *final state minus that of the initial state.* Mathematical abbreviations save time and space, and as a general rule we shall use Greek delta (Δ) to denote such a change, and Greek sigma (Σ) to denote 'sum of'. Thus on melting one mole of mercury, there is a volume change represented as follows.

	Hg(c) $\longrightarrow$	Hg(l)
Molar volume ($\bar{V}$)	14.134 ml	14.654 ml

$$\Delta\bar{V} = \Sigma\bar{V} \text{ (Products, final state)} - \Sigma\bar{V} \text{ (Reactants, initial state)}$$

$$= 14.654 - 14.134$$

$$= \mathbf{+0.520\ ml.}$$

The *positive* sign implies an *increase* of volume, whereas a negative sign would imply decrease. In quite general terms, for a change of a function of state, in which products (final state) are formed from reactants (initial state),

$$\Delta(\text{function of state}) = \sum\begin{pmatrix}\text{fn. of state,}\\ \text{products}\end{pmatrix} - \sum\begin{pmatrix}\text{fn. of state,}\\ \text{reactants}\end{pmatrix}. \quad (1.1)$$

Finally it must be stressed that to measure a function of state implies that equilibrium exists within the system. We can generalize only about systems which are thermally, mechanically, and chemically in equilibrium. This condition is seldom met. Slight differences of temperature are often found even in a carefully controlled apparatus, and the experimental thermodynamicist must be constantly alert to these possibilities.

1.2.3 Mechanical Work and Expanding Gases

When we take the lid off a bottle of cola, the released carbon dioxide *does work on the atmosphere.* That is, it has to push back the air, which resists the expansion, in order to make room for itself. The same is true of steam from water boiling in

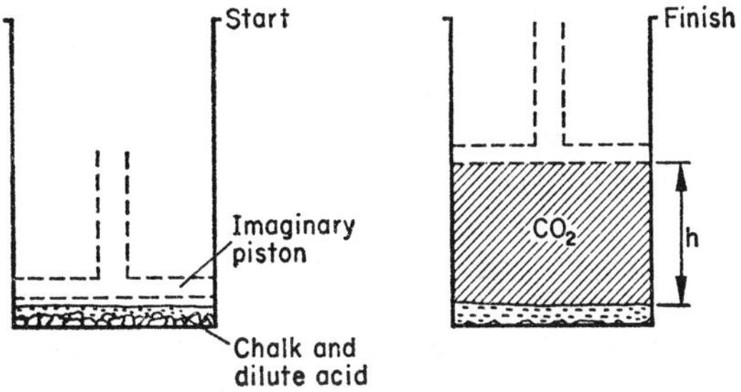

Fig. 1.1 Evolution of carbon dioxide causes work to be done on the gas in the surrounding atmosphere.

a kettle and of many chemical processes. This is one of the many forms of energy which we must learn to recognize. The expression we shall now derive will be quoted often.

Consider the reaction of dilute hydrochloric acid on calcium carbonate to be taking place in a tall form beaker. The evolved CO_2 will push back the air. It is convenient to imagine a weightless piston, which just fits the inside of the beaker, to be moving up at the same time. This is pictured in Fig. 1.1.

Mechanical work is equal to the product of the force and the distance moved, and in this case, for a small movement dh, work = f dh. If A is the cross-sectional area of the piston, and P the resisting external pressure (usually atmospheric), the total work

$$= \int PA\, dh = \int P\, dV = P\,\Delta V. \quad (1.2)$$

If we may assume that the gas behaves ideally, the general gas equation, $PV = nRT$ where n is the number of moles of gas, and R the universal gas constant can be used to simplify the measurements.

In passing it is as well to mention the numerical value of R. We know that a mole of gas at s.t.p. occupies 22.413 litres, and so, at 273.15 K, the equation of state becomes 1(22.413) = (1)R(273.15), which gives R = 0.082 05 litre atm/mol K.

Alternatively, we may express R in joule/mol K (1 litre atm = 101.3 joule), or in cal/mol K (4.184 joule = 1 (defined) calorie). To summarize these calculations,

$$\begin{aligned} R &= \mathbf{0.082\,05\ litre\ atm/mol\ K,} \\ &= \mathbf{8.314\,joule/mol\ K,} \quad \text{or} \\ &= \mathbf{1.987\,2\ cal/mol\ K} \end{aligned}$$

From time to time it will be necessary to use each one of these expressions.

1.2.4 The Absolute Temperature Scale

We all have a more or less well-developed sense of 'hot' and 'cold' and understand the concept of temperature to mean that high values are associated with 'hot' bodies, and low with 'cold'. The measurement of temperature on a numerical scale, however, has always presented problems. The basic difficulty is that the measured temperature *depends on the thermometer being used.*

The problem of the choice of thermometer is best understood by way of example. Let us begin by defining 100° Celsius as the temperature of boiling water at 1 atmosphere pressure, and 0°C as that of wet ice at 1 atmosphere in contact with air. We can then calibrate a mercury-in-glass thermometer by using these two points. Assuming a smooth capillary, we can interpolate a value of 50° exactly midway between the calibration points. We therefore define this intermediate temperature in terms of our chosen thermometer. However, if we construct thermometers using the e.m.f. of a thermocouple, or the volume of helium gas as the thermometric property, serious discrepancies arise. At 50 mercury degrees, the thermocouple reads 49.27°, and the gas thermometer 49.97°. *Is there any reason for choosing one thermometer as a standard, with which we can compare the others?* Before answering this question, we must pay some attention to the properties of gases.

All gases expand on heating. Thus a litre of nitrogen gas at 0°C expands by approximately 0.003 67 litre when heated by one degree. That is, if V_0 is the volume of the gas at 0°C, the volume at another temperature is given by

$$V_T = V_0(1 + \alpha T°\text{C})$$

where α is the coefficient of expansion, here equal to 0.003 67. The value of α is not absolutely constant, but varies slightly with the pressure of the gas. (See problem 1.1.) For an expansion from 0° to 1°C, we find that the ratio of volumes,

$$\frac{V_{(1°\text{C})}}{V_{(0°\text{C})}} = \frac{1/\alpha + 1}{1/\alpha}.$$

The term $1/\alpha$ has units of temperature. We are now able to define a new temperature scale such that $1/\alpha$ new degrees is equivalent to 0°C, and $(1/\alpha + 1)$ new degrees is equivalent to 1°C, and so on. This means that the gas volume is *directly proportional to temperature* measured on our new scale. In fact no gas obeys this relationship exactly, although helium comes very close to it. We none the less find that all gases obey it at the limit of zero pressure. The *hypothetical ideal gas* is thus defined as one which does obey it over all ranges. The temperature scale which results is named after Lord Kelvin (William Thomson) who was its originator.

The value of $1/\alpha$ for nitrogen at 1 atm is 272. Experiments with other gases indicate that the ice point, 0°C, is equivalent to a value of near 273 K. The

Kelvin scale is now defined with high accuracy such that the triple point of water (where ice, water and water vapour are all in equilibrium, at 0.01 °C) has the temperature 273.160 0° on the Kelvin scale. The triple point is more accurately defined than the ice point. On this basis **0 °C is 273.15 K.** Measurement on this ideal gas scale is best conducted with a constant pressure helium thermometer, although there are small deviations from the absolute scale. A comparison of the four temperature scales discussed above is given in Table 1.1.

TABLE 1.1 *Comparison between thermometers, calibrated linearly from* 0° *to* 100 °C *and the perfect gas scale*

Perfect gas scale, °C	Expansion of helium gas at constant pressure	Expansion of mercury in Jena glass 59	E.m.f. of chromel-alumel thermocouple
400	399.95	412.6	400.0
300	299.98	304.4	297.8
200	199.994	200.8	198.3
100	**100**	**100**	**100**
50	50.001	50.03	49.3
0	**0**	**0**	**0**
−50	−50.002		−45.4

1.3 Forms of Energy and Their Interconversion

We are today familiar with the idea that energy takes several forms. That is, we see in an electric motor, a dam, or an internal combustion engine, a common factor which we define as the capacity for doing work. The fact that the forms of the energy, electrical, gravitational, and chemical are fundamentally different does not detract from their working usefulness. Such understanding is of fairly recent origin, however. The idea that heat is a form of energy, manifesting itself as 'energy in transit' dates from 1842, when J. R. Mayer first published his hypothesis that heat and work were different forms of the same thing. On a voyage to the tropics, when bleeding a sick crew member, he noticed that the venous blood, usually dark and sluggish, was rich with oxygen and bright red in colour. This was because less oxygen was needed for oxidation of food to maintain the body temperature than in colder parts of the world. He then considered the possibility that the combustion of a given amount of food led to both heating and working of the body, but in different proportions depending on the need for each. The nature of heat, the 'caloric fluid', was an enigma up to that time, and even now it is a difficult concept. What Mayer came to realize, to use our terminology, was that *energy is conserved*, and that the primary chemical energy is converted into heat or work as the need demands, with no energy being lost in the process.

Since that time, we have become aware of new disguises for energy. Sometimes the *principle of the conservation of energy*, which is the simplest expression of the *First Law of Thermodynamics* has seemed to be invalidated, but in each case, the faith of those scientists who continued to believe in this great conservation law, was vindicated. A simple analogy will help to make the point.*

A young child had many toys, but among his favourites was a set of sixteen building bricks, with which he played almost every day. Each evening, his mother would start the job of collecting them and tidying them away ready for the next day. Sometimes they would be hidden, but the mother remembered how many there were, and she believed in the 'conservation of bricks' law, and so searched until they were found. One day it seemed that the child had put some bricks into his private lock-up box; it rattled. Although she could not see the bricks, she satisfied herself that there were in fact three bricks inside, by weighing the box, knowing its empty weight, and calculating that

$$(W_{\text{(box plus bricks)}} - W_{\text{(box)}})/W_{\text{(one brick)}} = \text{Number of bricks.}$$

On another occasion, the mother noticed that not only was the level of the gold-fish pond a little higher than normal, but also that the fish were showing visible signs of consternation. By making calculations based on the apparent change of volume caused by the bricks, the law of conservation of bricks was verified once more. In each case, her awareness of the bricks took a different form, even though the total number was unchanged.

So it is with energy. It wears different disguises, but each may be penetrated by appropriate mathematical or experimental study. As an example, we may mention the 'lost bricks' of radioactivity. The radioisotope of cerium, ^{144}Ce, for example, with a half-life of just over forty weeks, gives out atomic fragments and gamma rays at the rate of 25 watt/g. Where does the energy come from? This and many similar nuclear processes are explained in terms of change of mass, according to the Einstein equation, which relates energy to mass and the velocity of light, by $E = mc^2$. Another example concerned certain beta-decay processes, in which a true energy balance seemed not to be achieved. This was due to the emission of a further particle, the *neutrino* (christened thus, the 'little neutral one', by Enrico Fermi), which with its very small mass, and zero charge, was very difficult to detect.

Table 1.2 shows the seven main forms of energy, and some means of conversion between them. Only two-thirds of the possible types of energy converter are at present known. Those involving nuclear energy, the kind most recently developed, and gravitational, perhaps the oldest, have the lowest scores of all.

The inherent simplicity and efficiency of the fuel cell can be easily appreciated from the table. A conventional electricity generating station operates very ineffectively indeed. Starting with chemical energy in the form of fuel and oxygen, the first conversion from chemical to thermal energy entails an exothermic reaction in which heat is given out. The thermal energy is then converted to mechanical, via gaseous expansion, which runs the turbine in so doing. Finally, this energy is transmitted down the shaft to the generator, and electrical energy is at last available. Each of these energy-conversion steps is somewhat inefficient, and the overall efficiency is under fifty per cent. On the other hand, electrochemical cells or batteries, convert chemical energy directly to electrical, often with very high efficiencies. Fuel cells, which became well known as the auxiliary power supplies for the early manned space missions, are batteries in which the

* This is based on the story of Professor Richard Feynman, of the California Institute of Technology, with his permission.

TABLE 1.2 *Energy conversion chart*

FROM: \ TO:	Electrical	Gravitational	Chemical	Radiation	Thermal	Mechanical	Nuclear
Electrical		—	Electrolysis	Radio waves	Resistance heating	Motors	—
Gravitational	—		—	—	—	Falling weight	—
Chemical	Electrochemical cells	—		Chemiluminescence (Fireflies)	Exothermic reaction	Muscular conversion	—
Radiation	Photoelectric effect, radio aerial	—	Photo-synthesis		Radiant heating	Radiometer	γ-induced nuclear reactions
Thermal	Thermocouple, Thermoelectric effect	—	Endothermic reaction	Red-Hot poker		Gas expansion	—
Mechanical (Kinetic)	Generators, Piezo-electric effect	Rising weight	Compression in Haber reaction	Phosphorescence X-Ray production	Friction		New nuclei by bombardment
Nuclear	Emission of β-particles	—	—	Gamma-ray emission	Atomic fission	Atomic fission	

chemical fuels are continuously replenished. For journeys of medium duration, they offer the best energy-to-weight ratio, the cell and fuel being considered together as a unit. (For longer flights, a radioisotope generator will be preferable, because the weight of fuel for the fuel cell becomes excessive as distance increases.)

PROBLEMS

1.1 The coefficient of expansion for nitrogen gas at 0°C varies slightly with pressure as follows:

Pressure	(mmHg)	1000	760	400
α	(K^{-1})	$(3.674)10^{-3}$	$(3.670\,8)10^{-3}$	$(3.666)10^{-3}$.

Determine absolute zero in degrees Celsius as accurately as you can.

1.2 If 0.03 mole of $BaCO_3$ reacts with dilute acid at 20°C, calculate the work done by the evolved CO_2 on the atmosphere ($R = 8.314$ J/mol K).

1.3 Radioisotopes provide excellent low-power, long-life energy sources. The device SNAP-7B (System for Nuclear Auxiliary Power) powers an unattended lighthouse. It uses thermoelectric conversion; the hot junctions are clustered around the strontium-90 titanate source, which produces 0.45 watt/g.

(a) How many stages of energy conversion does this represent, and how does it compare with a conventional coal-burning power station with turbogenerators?

(b) If 1g of strontium-90 titanate is used for 5 years, to how many grams of fuel oil (calorific value = 43.93 kJ/g) is it equivalent? (Assume no diminution of power.)

2

The First Law of Thermodynamics

In this chapter, a mathematical expression of the first law will be discussed. The two functions *internal energy*, U, and *enthalpy*, H, will figure prominently. In addition, tabulations of standard enthalpy changes of formation, ΔH_f^0 for a number of compounds, will be given. From such tables it is possible to derive the amount of thermal energy associated with any reaction, as long as all reactants and products are listed.

2.1 Statement of the First Law

All matter contains energy. It contains energy in respect of atomic and molecular motion, of interatomic forces, of magnetic and electric fields in the nucleus as well as outside it, of its physical state, and in many other ways. In order to calculate the total energy contained in a sample we would first have to define a reference zero point. Such a zero would in the end be quite arbitrary. It is sufficient for us to consider only *changes* in the energy of a system, which, as we have seen, may be brought about principally by *thermal* or *mechanical* means. This and previous ideas are summarized in the formal statement of the First Law.

$$\boxed{\Delta U = U_2 - U_1 = q_{to} + w_{on}.} \tag{2.1}$$

The term ΔU represents the change of **internal energy** of the system, q_{to} is the thermal energy ('heat') *added to* the system, and w_{on} is the work *done on* the system. The subscripts 'to' and 'on' are emphasized here in order to avoid mistakes of sign in later calculations.

Internal energy U is a *function of state.* A change of internal energy from state 1 to state 2 must be considered only in terms of the initial and final states; *no route is specified.* The change from state 1 to state 2 may however be achieved in a multitude of different ways. (See Section 1.2.2.) Heat may be added, or work may be done, or heat and work may both be involved in different proportions. That is, q and w both depend on the route followed during a particular process, and are not functions of state. These ideas may be expressed diagrammatically as in Fig. 2.1. For a particular change, many routes are possible.

To use an analogy, one can travel from London to Manchester by train, by air, or even (with perseverance) by sea, but the geometric distance between them is fixed at 260 kilometres.

The work done on a system may be of many forms. For example, it might involve stretching of springs, lifting of weights, cutting of magnetic lines of force, or the compression of gases. In the chemical applications which follow, we shall be concerned almost exclusively with gaseous pressure-volume work.

If then we consider an infinitesimal expansion of a gaseous system, dV against an *external* pressure P_{ext}, the incremental work done *on the system* is, from eq. 1.2, $dw = -P_{ext}\, dV$. This becomes, for a significant amount of work, $w = \int P_{ext}\, dV$.

The early experiments in space travel served to emphasize that the first law applies in orbit as well as on land. Temperatures in the first space capsules were much too high for comfort, reaching 40°C or over for long periods at a time. This was due partly to the effect of solar radiation and partly to the chemical and metabolic reactions occurring inside the vehicle. For the time being, we shall overlook the effect of the sun. The space vehicle can be considered as a perfect example of a thermodynamic *system*; it has clearly defined boundaries and links with the surroundings are easily spotted. The body's normal processes of carbohydrate oxidation and so on are exothermic, and if the capsule were to be completely isolated the temperature of the interior would go on rising indefinitely. That is, both q and w would be zero, and $\Delta U = q_{to} + w_{on} = 0$ (isolated). In effect, chemical energy is converted to thermal energy, but none is lost; U is constant.

Because the capsule is surrounded by an almost perfect vacuum, heat loss by conduction is negligible. At this point the situation looks desperate, but in fact significant *radiation into space* can occur even at 300 K. That is, for a steady

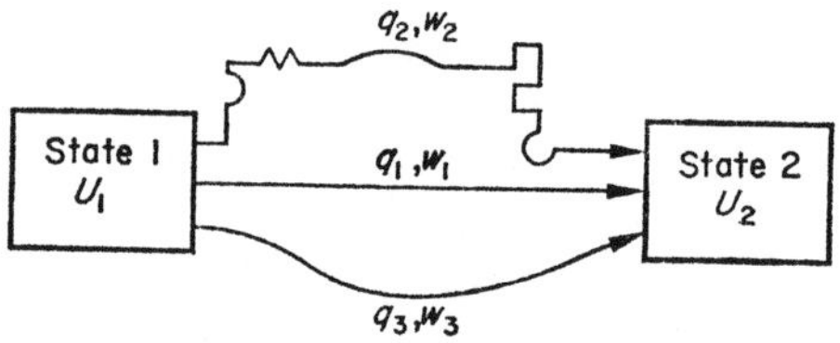

FIG. 2.1 Internal energy is a function of state only, although many routes exist between states 1 and 2. The change in U is the same, whether it is $\Delta U = q_1 + w_1$, or $= q_2 + w_2$, or $= q_3 + w_3$.

temperature to be maintained, all surplus thermal energy must be radiated away. What about the radiation from the sun, which we have conveniently forgotten up to now? This energy, once received, must be re-radiated as fast as it arrives. We can summarize the situation as we now see it. First, no work is done, so $w_{on} = 0$. Then,

$$q_{to} = q \text{ (from sun)} - q \text{ (radiated to space)}$$
$$= -\text{ve},$$

therefore

$$\Delta U = q_{to} + 0 = -\text{ve}.$$

That is, in order to maintain thermal equilibrium, *net* energy must leave the system.

In fact, the capsule is carefully oriented so that a larger surface is radiating to deep space than is absorbing radiation from the sun. Also, the emissivities of the surfaces may be carefully chosen to give the best conditions.

2.1.1 Reversible Expansion of an Ideal Gas

As the expansion of a gas proceeds, at a constant temperature, so its pressure drops. At any stage the amount of work it can do is determined by the opposing external pressure. If the opposing pressure is allowed to drop too quickly, less than the theoretical maximum work is done. If a railway engine develops wheel-slip, the work it does depends on the opposing frictional forces, not on its maximum (equilibrium) force. If a gas expands against an opposing pressure which is only *infinitesimally* less than its own, the term *reversible* is applied. (The two pressures are 'virtually equal'.) If the opposing pressure at any moment is increased infinitesimally, the direction of the process is reversed, and the gas is compressed again by a minute amount. Of course, such an overall process of compression or expansion would take an infinitely long time, and is never achieved in practice. It is none the less a useful theoretical model. In such a case, at constant temperature, the external pressure P_{ext} is *virtually equal* to P. Thus $w = -\int P\,dV$. For an ideal gas, $P = nRT/V$, and so

$$w = -nRT\int dV/V$$
$$= -nRT \ln V_2/V_1. \qquad (T \text{ constant}) \quad (2.2)$$

2.1.2 Constant Volume Processes

Many important experiments are carried out in closed containers, called bomb calorimeters (see Chapter 3). In this case the volume remains constant, and so $dV = 0$, and $w = -\int P\,dV = 0$. Forms of work other than that due to gaseous expansion are already assumed to be zero. The first law may then be written

$$\boxed{\Delta U = q_V}. \qquad (2.3)$$

The subscript V implies constant volume conditions. This means that the thermal energy added to a constant volume system is equal to the increase in internal energy. This affords a simple and useful means of measurement.

Example 2.1 In a typical experiment, 10 millimoles (0.01 mol) of hydrocarbon were burnt in an excess of gaseous oxygen in a bomb calorimeter, 21.97 kJ of heat were evolved. Thus $q_V = -2197$ kJ/mol for the process

$$\text{Hydrocarbon} + \text{oxygen} \longrightarrow x\ CO_2 + y\ H_2O,$$

and therefore $\Delta U = -2197$ kJ/mol. Note that because heat is evolved the internal energy decreases (ΔU has a negative sign). Such reactions are termed *exothermic*, while absorption of thermal energy at constant temperature characterizes an *endothermic* process.

2.1.3 Constant Pressure Processes

The vast majority of chemical and other processes take place at constant pressure, usually of one atmosphere. In these cases,

$$w = -\int P\,dV = -P\int dV = -P\Delta V$$

and $$\boxed{\Delta U = q_P - P\Delta V}. \qquad (\text{only } PV \text{ work, } P \text{ constant}) \quad (2.4)$$

Here, q_P is that heat added to the system during the constant pressure process. It is not necessary to qualify ΔU in this way (why not?).

2.2 A New Function—Enthalpy

We now define a new function, which is both important and supremely useful. This is the **enthalpy** of a system, which has the symbol H. An alternative name, often used, is that of *heat content*, which is preferable in so far as it is partly self descriptive. It can be confused with *heat capacity*, however, and so the term enthalpy will be used in this book. It is defined mathematically as

$$\boxed{H = U + PV}. \qquad (2.5)$$

Although it is difficult to visualize at this stage, it will become much more familiar with use. For the time being, we notice that, because both P and V are functions of state as well as U, H must itself be a function of state. That is, it describes the condition of a sample at a particular moment. Just as measurements of U in an absolute sense are quite inappropriate, so only changes in H are accessible to experiment.

Such an experiment or reaction would involve a 'before' state, 1, and an 'after' state, 2. We could write

$$H_2 = U_2 + P_2V_2,$$

and

$$H_1 = U_1 + P_1V_1;$$

on subtracting,

$$\Delta H = \Delta U + (P_2V_2 - P_1V_1). \qquad (2.6)$$

If we allow all these functions to vary at once, the analysis is unnecessarily complex, especially when we realize that most processes occur in the open at 1 atmosphere pressure. In these cases, $P_1 = P_2 = P$, say, and

$$\boxed{\Delta H = \Delta U + P\Delta V}. \qquad (P \text{ constant}) \ (2.7)$$

If we rewrite this as $\Delta U = \Delta H - P\Delta V$, and compare it with eq. 2.4 we can identify q_P as ΔH, that is

$$\boxed{\Delta H = q_P}. \qquad (\text{only } PV \text{ work, } P \text{ constant}) \qquad (2.8)$$

As in the case of ΔU, an exothermic change implies a negative sign for ΔH, an endothermic change implies a positive sign. Once again, it is stressed that q must always be qualified, because q_P is ΔH, although q_V is ΔU.

2.2.1 Relationship between ΔH and ΔU

Determinations of the calorific value (C.V.) of many gaseous fuels, and of all solid fuels, are performed in constant volume calorimeters. That is to say, values of $q_V = \Delta U$ are determined. When the fuel is burnt, however, at one atmosphere pressure, additional energy of expansion (+ve or −ve) against the atmosphere is involved, and the value of q actually realized, that is $q_P = \Delta H$ may be significantly different. We start with eq. 2.7,

$$\Delta H = \Delta U + P\Delta V.$$

The change of volume ΔV = (volume of products − volume of reactants). Combustion involves the gases oxygen and carbon dioxide, and perhaps gaseous fuels as well. If gases are involved in a reaction, they will account for by far the greatest part of the volume change, and we shall see that the volumes of solids and liquids are negligibly small in comparison. Suppose that we have n_1 moles of gas before the reaction, and n_2 moles of gas after it. Assuming the equation of state to be valid, we can write

$$PV_2 = n_2RT$$

and $$PV_1 = n_1RT$$

therefore $$P\Delta V = \Delta nRT.$$

Substituting into eq. 2.7, we have

$$\boxed{\Delta H = \Delta U + \Delta nRT}. \quad (2.9)$$

This equation is very useful, as will be seen in the following example.

Example 2.2 Consider the combustion of propane gas. The equation is

$$C_3H_8\,(g) + 5\,O_2\,(g) \longrightarrow 3\,CO_2\,(g) + 4\,H_2O\,(l).$$
$$n = 1 \qquad\quad 5 \qquad\quad 3$$

The letters g, l, c in parentheses denote the phase of each chemical to be gas, liquid or crystal. Measurement of $q_V = \Delta U$ for this reaction, with the initial and final temperature near 25 °C (298 K), the standard temperature for such measurements, gives $\Delta U = -2195$ kJ/mol of propane. Would more or less heat be liberated in a constant pressure combustion unit giving identical products (this implies that product steam is condensed)? Six moles of gaseous reactants give only 3 moles of gaseous product, so $\Delta n = (3 - 6) = -3$. That is, the atmosphere is *closing in on the system* while the process is occurring, and work is being done on the system. That is, additional energy is released (as heat) when constant pressure prevails. Using eq. 2.9, we find that

$$\begin{aligned}\Delta H &= \Delta U + \Delta nRT \\ &= -2195 + (-3)(8.314)(298)/(1000) \\ &= -2195 - 7.3 = \mathbf{-2202\ kJ/mol.}\end{aligned}$$

Thus ΔH is in fact more negative than ΔU, although the difference is a very small proportion of the whole. (Moreover, the volume of water, $= 4 \times 18 = 72$ ml, is truly insignificant compared with $3(25) = 75$ litres, the approximate volume change of the gases.) For all that, the difference between ΔH and ΔU is frequently several per cent, when for example ΔH itself is small, or the temperature is high.

2.3 Uses and Conventions of ΔH

2.3.1 Enthalpy Change of Reaction

During the building of the Aswan High Dam many thousands of tons of concrete were poured. The primary setting process, that of hydration of the aluminosilicates, is exothermic. To avoid overheating and cracking, many miles of water-cooling pipes are embedded in the concrete. The reaction, and enthalpy change, may be represented as

$$\text{Cement} + \text{water} \longrightarrow \text{Concrete}, \ \Delta H = -28.9\ \text{kJ/mol}\ H_2O.$$

That is, at constant pressure, 28.9 kJ of thermal energy is evolved per mole of water. The value of ΔH must be known before any reaction is carried out on a large scale, either industrially or in a laboratory. The heat must be removed (by heat-exchanger or reflux condenser as appropriate) because overheating could allow a reaction to get out of hand. Some Grignard reactions can occur dangerously fast if the temperature much exceeds 10°C.

Combustion processes constitute an important class of reactions, and the enthalpy changes associated with them give important information for operators of industrial plant as well as for internal combustion engineers. As an example, consider the hydrocarbon 3,3-dimethyl pentane, C_7H_{16}, a component of high-octane petrol. For combustion to carbon dioxide and liquid water,

$$C_7H_{16}\,(l) + 11\ O_2\,(g) \rightarrow 7\ CO_2\,(g) + 8\ H_2O\,(l),$$

ΔH is -4803 kJ/mol. Such a value would originate from an experiment with a constant volume calorimeter, using the method of Section 2.2.1 to convert ΔU to ΔH. The negative sign implies exothermicity.

Acid-base neutralization occurs exothermically with the formation of dissociated salt, and water:

$$Na^+ + OH^- + H_3O^+ + Cl^- \rightarrow Na^+Cl^- + 2\ H_2O.$$

In the case of a strong acid and a strong base, not only the salt, but also the reactants are completely dissociated. The reaction is effectively that between hydroxonium and hydroxyl ions, and it will apply for *any* strong acid with *any* strong base. As a result, we find that the *enthalpy change of neutralization* is approximately constant for any such reaction, and equals -56.07 kJ/mol.

Not only chemical reactions, but *physical changes* may be assigned values of ΔH. Thus a vaporization process, for example,

$$Br_2(l,\ 59°C,\ 1\ atm) \rightarrow Br_2(g,\ 59°C,\ 1\ atm)$$

is assigned a value for ΔH of $+29.2$ kJ/mol. Thus under these constant pressure conditions the enthalpy of bromine vapour is greater than that of the liquid by 29.2 kJ/mol. The alternative name of '*heat content*' for H is particularly easy to accept in this instance, as the vapour clearly contains 'more heat'. Not only vaporization, but fusion, sublimation, solid/solid phase changes, and also nuclear reactions may be described in terms of ΔH.

2.3.2 Standard Enthalpies of Formation

It will be clear from what has been said that there are multitudes of possible reactions which might concern us, and therefore multitudes of ΔH values. New compounds are being synthesized daily, and each will undergo many possible reactions. Clearly, it is almost impossible to determine ΔH for every possible change. Luckily, it is also unnecessary. It is sufficient, as we shall see, to know the **standard enthalpy change of formation** (sometimes abbreviated to *enthalpy of formation* or called *standard heat of formation*), for each compound with which we are dealing. The formation reaction is that in which the compound is formed from its elements in their **standard states,** that is, their most stable form under the chosen *standard conditions* of temperature and pressure. These are usually taken to be one atmosphere pressure and a temperature of 25°C (298.15 K). Under these conditions, for example, iodine is normally crystalline, hydrogen is gaseous, and sulphur exists as the rhombic crystalline form. In general, the compound formed will also be in its standard state. (This is not always the case, however. Thus it is possible, by a form of extrapolation, to calculate the standard enthalpy of formation of water vapour at 1 atm and 25°C, although it is normally liquid at this temperature. It is convenient to use such values for reactions at temperatures higher than 100°C, say, because allowance will already have been made for the heat of vaporization of the water.) The omnibus expression for the enthalpy of formation is shown below.

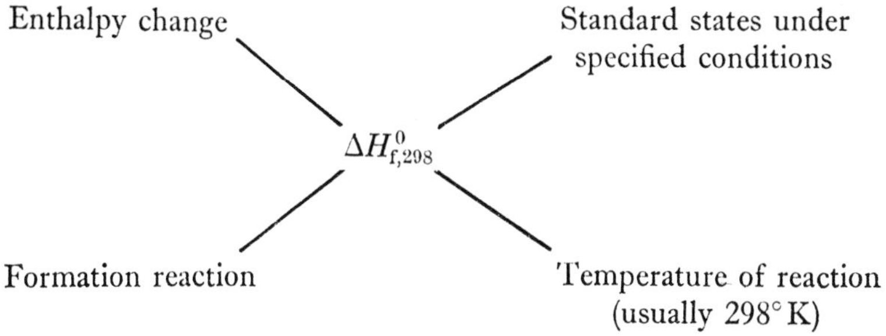

Unless specified otherwise, $\Delta H^0_{f,T}$ where T is not 25°C, will refer to the reaction in which both reactants and product are in their standard states at the stated temperature, T.

If aqueous solutions are involved, the standard state implies unit *effective* concentration, usually denoted 'aq'. (This topic will be more fully discussed in Chapter 7.) Standard enthalpies of formation of salt solutions may be determined, and when ΔH^0_f for the hydrogen ion is taken *arbitrarily* to be zero, the values for individual ions may be quoted. In effect, we regard the hydrogen ion as an element. This is necessary because anions cannot be studied in isolation from cations.

Table 2.1 shows selected values of $\Delta H^0_{f,298}$, for a range of compounds, free atoms, and hydrated ions. Values are quoted in joules or kilojoules. However, values of ΔH^0_f are only known to the nearest 50 joules at best, and to 500 or more joules in many cases. (Values of $\Delta G^0_{f,298}$, and of S^0_{298} are also shown, and these will be discussed in later chapters.) It should be noted that all values of ΔH^0_f for all elements are *zero*; this is a necessary consequence of the definition.

Example 2.3 The standard enthalpy change of formation for crystalline Na_2O is found to be -415.9 kJ/mol. This necessarily implies the following exact equation.

$$2\,Na\,(c) + \tfrac{1}{2}\,O_2\,(g) \longrightarrow Na_2O\,(c),\ \Delta H^0_{f,298} = -415.9\ kJ.$$

Just over four hundred kilojoules are evolved per mole of crystalline sodium oxide formed. Sodium is normally solid, and oxygen gaseous under the standard conditions, and so the starting materials and product are completely specified.

Example 2.4 The enthalpy of formation of crystalline $MgBr_2$ is

$$\Delta H^0_{f,298}(MgBr_2\,(c)) = -517.6\ kJ/mol.$$

The temperature of 298 K implies the physical states shown in the equation

$$Mg\,(c) + Br_2\,(l) \longrightarrow MgBr_2\,(c).$$

However, it is quite possible that information is required at, say, 373 K, at which temperature the vapour pressure of bromine is well over 1 atmosphere. Therefore the standard state of bromine is now the gas, and the relevant equation becomes

$$Mg\,(c) + Br_2\,(g) \longrightarrow MgBr_2\,(c),\ \text{with}\ \Delta H^0_{f,373} = -548.1\ kJ/mol.$$

Such a statement assumes the knowledge that bromine boils at 59°C, and the

TABLE 2.1 *A selection of basic thermodynamic data*[a]

Substance	State[b]	$\Delta H^0_{f,298}$ kJ/mol	$\Delta G^0_{f,298}$ kJ/mol	S^0_{298} J/mol K
Aluminium				
Al	g	326	285.8	164.4
Al	c	0	0	28.3
Al_2O_3	c	−1676	−1582	50.9
Barium				
Ba	g	175.6	144.8	170.3
Ba	c	0	0	67
Ba^{2+}	aq	−538.4	−560.7	13
$BaCl_2$	c	−860.1	−810.9	126
$BaSO_4$	c	−1465	−1353	132.2
Boron				
B	g	562.7	518.8	153.3
B	c	0	0	5.86
B	amorphous	3.77	3.56	6.53
BF_3	g	−1137	−1120	254.0
B_2H_6	g	35.56	86.6	232.0
Bromine				
Br	g	111.9	82.43	174.9
Br^-	aq	−121.5	−104	82.4
Br_2	l	0	0	152.2
HBr	g	−36.40	−53.43	198.6
HBr	aq	−121.5	−104	82.4
Calcium				
Ca	g	192.6	158.9	154.8
Ca	c	0	0	41.6
Ca^{2+}	aq	−543.0	−553	−55.2
$CaCO_3$	c	−1206	−1129	92.9
$CaCl_2$	c	−795	−750.2	113.8
CaO	c	−635.6	−604.2	39.7
$Ca(OH)_2$	c	−986.6	−896.8	76.1
Carbon				
C	g	716.7	671.3	158.0
C	graphite	0	0	5.7
CO_2	g	−393.5	−394.4	213.6
CO	g	−110.5	−137.2	197.6
CS_2	l	89.70	65.27	151.3
CS_2	g	117.4	67.15	237.7
Chlorine				
Cl	g	121.7	105.7	165.1
Cl^-	aq	−167.2	−131.3	56.5
Cl_2	g	0	0	223.0
HCl	g	−92.30	−95.27	186.8
HCl	aq	−167.2	−131.3	56.5

Substance	State[b]	$\Delta H^0_{f,298}$ kJ/mol	$\Delta G^0_{f,298}$ kJ/mol	S^0_{298} J/mol K
Copper				
Cu	g	341.1	301.4	166.3
Cu	c	0	0	33.3
Cu^{2+}	aq	64.39	64.98	−98.7
CuO	c	−155.2	−127.2	43.5
Cu_2O	c	−166.7	−146.4	100.8
$CuSO_4$	c	−770	−661.9	113.4
$CuSO_4.H_2O$	c	−1084	−917.1	149.8
Fluorine				
F	g	78.99	61.59	158.6
F_2	g	0	0	202.7
HF	g	−271.1	−273.2	173.7
HF	aq	−332.6	−278.8	−13.8
Hydrogen				
H	g	218.0	203.3	114.6
H^+	aq	0	0	0
H_2	g	0	0	130.6
H_2O	g	−241.8	−228.6	188.7
H_2O	l	−285.8	−237.2	69.9
H_2O_2	l	−187.8	−120.4	109.6
Iodine				
I	g	106.8	70.29	180.7
I^-	aq	−55.19	−51.59	111.3
I_2	c	0	0	116.1
I_2	g	62.43	19.37	260.6
HI	g	26.48	2.13	206.5
HI	aq	−55.19	−51.59	111.3
Iron				
Fe	c	0	0	27.2
Fe^{2+}	aq	−87.86	−84.94	−113.4
FeO	c	−266.5	−244.3	54.0
Fe_2O_3	c	−822.2	−741.0	90.0
Magnesium				
Mg	c	0	0	32.5
$MgCl_2$	c	−641.8	−592.3	89.5
MgO	c	−601.8	−569.6	26.8
Nickel				
Ni	c	0	0	30.1
NiO	c	−244.3	−216.3	38.6
Nitrogen				
N	g	472.7	455.6	153.2
N_2	g	0	0	191.5
NH_3	g	−46.11	−16.48	192.3
NO	g	90.25	86.57	210.7
N_2O	g	82.05	104.2	219.7

Substance	State[b]	$\Delta H^0_{f,298}$ kJ/mol	$\Delta G^0_{f,298}$ kJ/mol	S^0_{298} J/mol K
Nitrogen (*cont.*)				
NO_2	g	33.18	51.30	240.0
N_2O_4	g	9.16	97.82	304.2
Oxygen				
O	g	242.2	231.8	161.0
O_2	g	0	0	205.0
O_3	g	142.7	163.2	238.8
Phosphorus				
P	c, white	0	0	41.1
PCl_3	g	−287.0	−267.8	311.7
PCl_5	g	−374.9	−305.4	364.5
Potassium				
K	c	0	0	64.2
K^+	aq	−252.4	−283.3	102.5
Silicon				
Si	c	0	0	18.8
SiH_4	g	34.31	56.9	204.5
SiO_2	c	−910.9	−856.7	41.8
Silver				
Ag	c	0	0	42.7
AgCl	c	−127.1	−109.8	96.2
$AgClO_2$	c	0·0	66.94	134.6
Sodium				
Na	c	0	0	51.2
Na^+	aq	−240.1	−261.9	59.0
NaCl	c	−411.0	−384.0	72.4
Na_2O	c	−415.9	−376.6	72.8
NaI	c	−287.9	(−268.4)[c]	(102.5)
Sulphur				
S	g	278.8	238.3	167.7
S	rhombic c	0	0	31.8
S_2	g	128.4	79.08	228.1
$SOCl_2$	l	−245.6	−203.4	215.7
H_2S	g	−20.63	−33.56	205.7
SO_2	g	−296.8	−300.2	248.1
SO_4^{2-}	aq	−909.3	−744.6	20.1
Titanium				
Ti	c	0	0	30.3
$TiCl_4$	l	−750.2	−674.5	252.7
TiO_2	c	−912.1	−852.8	50.2
Tungsten				
W	c	0	0	33.5
WI_2	g	(−8.37)[c]	(−38.45)	(251)

Substance	State[b]	$\Delta H^0_{f,298}$ kJ/mol	$\Delta G^0_{f,298}$ kJ/mol	S^0_{298} J/mol K
Zinc				
Zn	c	0	0	41.6
Zn^{2+}	aq	−152.4	−147.2	−106.5
ZnO	c	−348.0	−318.2	43.9
ZnS	c	−202.9	−198.3	57.7
Organic compounds				
CH_4	g	−74.81	−50.75	186.1
C_2H_2	g	226.7	209.2	200.8
C_2H_4	g	52.28	68.12	219.5
C_2H_6	g	−84.67	−32.89	229.5
C_3H_8	g	−103.8	−23.49	269.9
n-C_4H_{10}	g	−124.7	−15.71	310.0
iso-C_4H_{10}	g	−131.6	−17.97	294.6
C_6H_6	l	82.93	129.7	269.2
CCl_4	l	−135.4	−65.27	216.4
$CHCl_3$	l	−134.5	−73.72	201.7
CH_3Cl	g	−80.83	−57.40	234.5
HCHO	g	−117.2	−113.0	218.8
CH_3OH	g	−200.7	−162.0	239.7
CH_3OH	l	−238.7	−166.4	126.8
C_2H_5OH	l	−277.6	−174.8	160.7
C_2H_5OH	g	−235.3	−168.6	282.0
CH_3COOH	l	−487.0	−392.5	159.8
HCN	g	130.5	120.1	201.8

[a] Data are taken from two publications of the National Bureau of Standards, Washington, D.C. entitled *Selected Values of Chemical Thermodynamic Properties*. Data for compounds of Groups I and II, and of the *d*-block of the periodic table are taken from Rossini, F. D., *et al.*, *Circular 500*, 1952. Data for compounds containing *only* elements of the *p*-block, including the rare gases, and hydrogen, are contained in *Technical Note 270*, Parts 1(1965), 2(1966) by Wagman, D. D. *et al.*, which is a continuing publication.

[b] g—hypothetical ideal gas at unit fugacity; aq—hypothetical ideal solution of unit molality; l, c—pure liquid or crystal at one atmosphere pressure.

[c] Parentheses indicate an estimated value, which is *not* taken from reference *a*.

difference of approximately 30.5 kJ accounts for the heat of vaporization which is involved. Notice also that formation of the compound at 373 K requires that the Mg (c) be at 373 K, and that the heat of formation of Mg (c) at 373° *is still zero* at this 'new' temperature.

Example 2.5 The enthalpy of formation of barium ions is found to be −538.4 kJ/mol. We saw that these are to be formed from elements or hydrogen ions, and such a reaction is given by the complete equation

$$Ba\ (c) + 2\ HCl\ (aq) \rightarrow H_2\ (g) + BaCl_2\ (aq),$$

which, because the ions are completely dissociated, may be written in the simplified form,

$$Ba\ (c) + 2\ H^+\ (aq) \rightarrow H_2\ (g) + Ba^{2+}\ (aq).$$

Apart from Ba^{2+} ions, every species is regarded as an element, with zero enthalpies of formation. It is therefore this equation to which the datum refers.

2.3.2 The Many Uses of ΔH_f^0 Data

We shall now see how the basic formation data may be used to provide information on a host of other reactions.

Two of the many possible chemical routes by which carbon may be converted to carbon dioxide are represented by the following equations.

Route I $\quad$ C (graphite) + O_2 (g) $\rightarrow$ CO_2 (g).

Route IIa $\quad$ C (graphite) + $\frac{1}{2}$ O_2 (g) $\rightarrow$ CO (g), $\quad$ followed by

b $\quad$ CO (g) + $\frac{1}{2}$ O_2 (g) $\rightarrow$ CO_2 (g).

It is clear that the initial state (reactants, carbon plus oxygen), and the final state (product, carbon dioxide gas) are identical for both routes. When enthalpy was introduced, it was agreed that it was a function of state, such that changes of it *did not depend on route.* In the present example, it follows that ΔH for Route I and that for Route II must be identical. That is to say that the overall change in H is equal to the sum of the values of ΔH for each step. Thus,

$$\Delta H\,(\mathrm{I}) = \Delta H\,(\mathrm{IIa}) + \Delta H\,(\mathrm{IIb}).$$

Now equation I represents formation of CO_2, and the equation IIa formation of CO. Values are

$$\Delta H_f^0(\mathrm{CO_2\,(g)}) = -393.5\ \mathrm{kJ/mol}.$$
$$\Delta H_f^0(\mathrm{CO\,(g)}) = -110.5\ \mathrm{kJ/mol}.$$

Thus
$$(-393.5) = (-110.5) + \Delta H(\mathrm{IIb}),$$
$$\Delta H(\mathrm{IIb}) = \mathbf{-283.0\ kJ/mol.}$$

That is, the combustion of CO to give gaseous CO_2 is accompanied by evolution of 283 kJ/mol.

This, the **principle of constant heat summation,** often known as **Hess's Law,** is thus seen to lead directly from the fact that H is a function of state. This idea is immensely powerful, because it enables ΔH_{298}^0 values to be determined for any reaction, as long as the enthalpies of formation are known for each reactant and product.

Consider an equation of the general form

$$\mathrm{A} + \mathrm{B} \ldots \rightarrow \mathrm{P} + \mathrm{Q} \ldots$$

for which we wish to know ΔH^0. Imagine this to take place in two stages, first the 'unformation' of A, B, etc. into their elements, then the formation of P, Q, etc. from those same elements. That is A + B ... $\xrightarrow{\Delta H_1^0}$ (elements in standard states) $\xrightarrow{\Delta H_2^0}$ P + Q ...

Now $\quad \Delta H_1^0 = -(\Delta H_f^0(\mathrm{A}) + \Delta H_f^0(\mathrm{B}) + \ldots)$ (unformation)

and $\quad \Delta H_2^0 = \ (\Delta H_f^0(\mathrm{P}) + \Delta H_f^0(\mathrm{Q}) + \ldots)$ (formation).

Thus, for the overall reaction,

$$\Delta H^0 = \Delta H_1^0 + \Delta H_2^0$$
$$= (\Delta H_f^0(\mathrm{P}) + \Delta H_f^0(\mathrm{Q})) - (\Delta H_f^0(\mathrm{A}) + \Delta H_f^0(\mathrm{B})).$$

This may be written in the abbreviated form

$$\boxed{\Delta H^0 = \Sigma(\Delta H_f^0(\text{Products})) - \Sigma(\Delta H_f^0(\text{Reactants}))} . \qquad (2.10)$$

This general equation is seen to fit the prototype eq. 1.1, and will be much used.

Example 2.6 Fluorine has been considered for use as an oxidant in rocket propellants. Its high reactivity ensures high energy release, and consequently high thrust. Its reaction with ammonia,

$$NH_3\,(g) + \tfrac{3}{2}\,F_2\,(g) \longrightarrow \tfrac{1}{2}\,N_2\,(g) + 3\,HF\,(g)$$

is being investigated. Let us calculate ΔH^0_{298} for this reaction.

Ammonia is being decomposed and hydrogen fluoride produced. Fluorine and nitrogen are elements in their standard states, and so have zero enthalpies of formation. By using data from table 2.1, and eq. 2.10, we may write

$$\begin{aligned}\Delta H^0_{298} &= \tfrac{1}{2}\,\Delta H_f^0(N_2) + 3\,\Delta H_f^0(HF) - \Delta H_f^0\,(NH_3) - \tfrac{3}{2}\,\Delta H_f^0(F_2)\\ &= 0 \qquad + 3(-271.1) - (-46.1) \qquad - 0\\ &= \mathbf{-767.3\ kJ/mol\ NH_3.}\end{aligned}$$

This represents a very high energy yield, especially when based on the mass of (fuel + oxidant). This energy is supplied by only $17 + \tfrac{3}{2}(38) = 74$ g of reactants (See Problem 3.7.)

Example 2.7 What heat is evolved on precipitation of one mole of $BaSO_4$ from solution of unit concentration?

Such a reaction could be represented as

$$Ba^{2+}\,(aq) + SO_4{}^{2-}\,(aq) \longrightarrow BaSO_4\,(c).$$

In this case, we find that

$$\begin{aligned}\Delta H^0_{298} &= \Delta H_f^0(BaSO_4) - \Delta H_f^0(Ba^{2+}) - \Delta H_f^0(SO_4{}^{2-})\\ &= -1465 \qquad - (-538.4) \quad - (-909.3)\\ &= \mathbf{-17.3\ kJ/mol.}\end{aligned}$$

Example 2.8 This example is taken from a paper by Margaret A. Frisch and J. L. Margrave [1] of the University of Wisconsin. They measured ΔH^0_{298} for the reaction

$$NO\,(g) + CO\,(g) \longrightarrow \tfrac{1}{2}\,N_2\,(g) + CO_2\,(g)$$

calorimetrically, and found a value of -373.2 kJ/mol CO_2. It is required to use this result to check the value for the enthalpy of formation of nitric oxide.

It is clear that if values for ΔH_f^0 are available for CO and CO_2, this can be done by using eq. 2.10. Thus

$$\begin{aligned}\Delta H^0_{298} &= \tfrac{1}{2}\,\Delta H_f^0(N_2) + \Delta H_f^0(CO_2) - \Delta H_f^0(NO) - \Delta H_f^0(CO)\\ -373.2 &= 0 \qquad + (-393.5) - \Delta H_f^0(NO) - (-110.5)\end{aligned}$$

whence $\quad \mathbf{\Delta H_f^0(NO) = +90.21\ kJ/mol.}$

This is the currently accepted best value for nitric oxide. Notice that the formation reaction is endothermic.

An alternative, and equally useful form of Hess's Law says in effect that *if simple reaction equations are added (or subtracted) to give an overall reaction* *scheme, then so may the corresponding ΔH values be added (or subtracted) to give an overall value of enthalpy change.* One final example will be given to demonstrate this alternative approach.

Example 2.9 The enthalpy changes for the following three reactions, which represent the formation of dilute boric acid solution, the dissolution of crystalline boric oxide, and the formation of liquid water respectively, are as follows:

I $2\,B\,(c) + 3\,H_2\,(g) + 3\,O_2\,(g) + aq \longrightarrow 2\,H_3BO_3$ (dil. solution)
$\Delta H^0 = -2153$ kJ/mol,

II $B_2O_3\,(c) + 3\,H_2O\,(l) + aq \longrightarrow 2\,H_3BO_3$ (dil. solution)
$\Delta H^0 = -17.4$ kJ/mol,

III $H_2\,(g) + \frac{1}{2}\,O_2\,(g) \longrightarrow H_2O\,(l)$ $\Delta H^0 = -285.9$ kJ/mol.

The term aq by itself represents an excess of solvent water. It is required to find ΔH for the reaction

$$2\,B\,(c) + \tfrac{3}{2}\,O_2\,(g) \longrightarrow B_2O_3\,(c), \tag{2.11}$$

which represents the formation of boric oxide from its elements.

Problems such as these may be solved by combining equations in various ways, by observation, until an acceptable route for the required reaction is found. Here, one possible approach would be to observe that reaction 2.11 has as reactant 2 B (c), and so does reaction I. The unwanted product of I, dilute boric acid solution, may be eliminated by performing II *in a reverse direction*; we note in passing that this involves an unwanted product of 3 H_2O (l), produced chemically, and not as solvent. This may in turn be dealt with by performing III, in reverse, three times over. At this stage we can write down the possible recipe I–II–3(III), which in full is:

$$2\,B\,(c) + 3\,\cancel{H_2\,(g)} + 3\,O_2\,(g) + aq \longrightarrow 2\,\cancel{H_3BO_3}\ (\text{dil. solution})$$
$$2\,\cancel{H_3BO_3}\ (\text{dil. solution}) \longrightarrow B_2O_3\,(c) + 3\,\cancel{H_2O\,(l)} + aq$$
$$3\,\cancel{H_2O}\,(l) \longrightarrow 3\,\cancel{H_2\,(g)} + \tfrac{3}{2}\,O_2\,(g).$$

Notice that many items are both produced and consumed, and undergo no net change, and so may be deleted. On adding up these three reactions, we are presented with the desired process, 2.11. (If we were not lucky the first time, we would learn to see which further reaction was needed as ingredient in the mixture.) The crux of the matter is this; having followed the recipe I–II–3(III) for the equations, *ΔH values may be treated in an identical manner.*

Thus
$$\Delta H = \Delta H_{I} - \Delta H_{II} - 3\,(\Delta H_{III})$$
$$= -2153 - (-17.4) - 3\,(-285.9)$$
$$= \mathbf{-1278\ kJ/mol\ B_2O_3}.$$

In these few examples, we have seen how very straightforward is the determination of ΔH for a reaction, either from formation data, or from related reactions. This is a skill which should become second nature to a student of thermodynamics, because such techniques also apply to free energy changes, ΔG^0_{298} (Ch. 6), which can give invaluable information on chemical and other equilibria.

REFERENCES

[1] FRISCH, MARGARET A. and MARGRAVE, J. L. *J. phys. Chem.*, **69**, 3863 (1965).

PROBLEMS

2.1 The decomposition of krypton difluoride

KrF_2 (g) $\longrightarrow$ Kr (g) + F_2 (g) has been studied at 25°C by S. R. Gunn (*J. Am. chem. Soc.* **88**, 5924 (1966)), who found that $\Delta U = -59.4$ kJ/mol.
(a) Calculate ΔH for this reaction.
(b) For sublimation of KrF_2 (c), $\Delta H^0_{sub} = +41.4$ kJ/mol. Estimate $\Delta H_{f,298}$ (KrF_2 (c)).

2.2 Gaseous normal propyl chloride may be hydrogenated to form propane and HCl gas; ΔH^0_{373} is -65.7 kJ/mol. The related reaction with iso-propyl chloride occurs with $\Delta H^0_{373} = -58.37$ kJ/mol. What is ΔH^0_{373} for the reaction

$$\text{iso-}C_3H_7Cl \longrightarrow \text{n-}C_3H_7Cl?$$

2.3 In order to determine the enthalpy of formation of CH_3Br more accurately, ΔH^0_{298} for the reaction

$$CH_3Br\ (g) + H_2\ (g) \longrightarrow CH_4\ (g) + HBr\ (g)$$

has been determined as -73.55 kJ/mol. Use data from Table 2.1 to assist you in bringing this project to its conclusion.

2.4 Enthalpies of combustion for the different liquid isomers of hexane have been accurately determined, as follows:

Isomer	ΔH^0_{comb}, kJ/mol
n-hexane	4141.3
2-methyl pentane	4135.9
3-methyl pentane	4138.2
2,3-dimethyl butane	4133.2
2,2-dimethyl butane	4126.9

What are the enthalpy changes for isomerization, for the reaction

$$\text{n-hexane} \longrightarrow \text{isomer},$$

in each case?

2.5 Diborane has been considered as a high-energy rocket fuel. Assuming it burns to HBO_2 (c) and H_2O (l), calculate the thermal energy available per gram of fuel-oxygen mixture. (This may be compared with a value of 8.24 kJ/g mixture for n-hexane.) For HBO_2 (c), $\Delta H^0_f = -788.7$ kJ/mol; other data from Table 2.1.

2.6 Acetic acid is currently manufactured by a high temperature, high pressure reaction:

$$CH_3OH\ (g) + CO\ (g) \longrightarrow CH_3COOH\ (g).$$

Enthalpies of vaporization are 35.23 (methanol) and 24.31 (acetic acid) kJ/mol. Using data from Table 2.1, estimate ΔH for this reaction.

2.7 Delrin (DuPont TM) is a stabilized poly-formaldehyde. The monomer is produced by partial oxidation of methanol to HCHO and H_2O, at 600°C. What is ΔH^0_{298}? The value of ΔH under plant conditions is -154.0 kJ/mol. Why is the agreement not exact?

2.8 The compound $CHClF_2$, an intermediate in the production of PTFE is produced by the reaction

$$CHCl_3 + 2HF \rightarrow CHClF_2 + 2HCl.$$

Is this reaction endothermic? ($\Delta H^0_f(CHClF_2) = -491.6$ kJ/mol.)

2.9 Use Table 2.1 to assess the usefulness of propane and n-butane as fuels. Gram for gram, which provides more heat when burnt to CO_2 and liquid H_2O? (H = 1.008, C = 12.01.)

2.10 Vaughn and Muetterties (*J. phys. Chem.* **64**, 1787 (1960)) have determined $\Delta H = -413.0$ kJ/mol SF_4 for the reaction

$$SF_4\,(g) + 2H_2\,(g) \rightarrow S\,(c) + \frac{4}{n}(HF)_n\,(l).$$

This liquid HF product is partly polymerized, but it is known that the enthalpy of formation of $\frac{1}{n}(HF)_n$ when equally polymerized is $\Delta H_f = -282.8$ kJ/20 g HF (Mol Wt of HF monomer = 20.0). Determine $\Delta H_f(SF_4\,(g))$.

3

Thermochemistry

In this chapter we shall first look in more detail at the various experimental *methods of measurement.* These include determinations not only of heats of reaction, as ΔH or ΔU, but also of the *heat capacities* of individual materials. Armed with such information, we shall see that if the value of ΔH for a reaction is known at one temperature, it is possible to calculate it at another temperature. Finally, by using the concept of *bond energies,* we shall learn how to estimate values of ΔH_f for compounds which are rare or uncharacterized.

3.1 Calorimetry

This is one of the best established of the experimental sciences, and the ever-present challenge of more accurate determinations continues to draw the attention of research workers. Although the art and science of calorimetry is so well established, basically new techniques are being developed almost month by month. Because reactions vary so much, with respect to reaction rates, physical states, corrosive qualities and so on, there are very many types of calorimeter. However, they fall into a small number of broad classes. Because some calorimeters give ΔH directly, while others yield ΔU data, a short summary of types will follow.

3.1.1 Bomb Calorimeters

This type of apparatus is by far the most often used. The reactants, previously sealed in a strong container, are allowed to react under constant volume conditions. Gases at high pressures are often used, hence the name. A line diagram of such an apparatus is shown in Fig. 3.1.

The sequence of events during a typical experiment, in which an oxide is formed from a solid element and excess gas, is as follows. The element is weighed carefully, put in place, the igniting wire is positioned, and the bomb sealed tightly. Oxygen is then pumped in at high pressure. When the temperature, T_1, of the system which includes the water jacket has been measured, a small electric current is passed to initiate the reaction. (The energy introduced through the fuse wire is compensated for later.) The temperature rise of the calorimeter, which is usually only a degree or two, is measured as accurately as possible. (Final temperature $= T_2$.) Then, in a separate control experiment, the products plus calorimeter are raised from T_1 to T_2 by *electrical heating,* and the amount of electrical energy used is carefully measured. Thus, in the first case,

$$\text{(A) System with reactants } (T_1) \longrightarrow \text{System with products } (T_2),$$

and $q = w = 0$, so $\Delta U_A = q + w = 0$.

In the second case, electrical heating introduces *Eit* joules (*E* volts, *i* amps, flowing for *t* seconds), and so for

(B) System with products (T_1) $\longrightarrow$ System with products (T_2),

$q = Eit$, $w = 0$, and so $\Delta U_B = Eit$.
Because *U* is a function of state, we can follow process (A) with the reverse of (B), giving

System with reactants (T_1) $\longrightarrow$ System with products (T_1),

for which

$$\Delta U = \Delta U_A - \Delta U_B$$
$$= -Eit \text{ J}.$$

That is to say that the electrical energy is equivalent to the energy 'released' by the reaction when carried out at constant temperature T_1. By this means we may refer to ΔU at *constant* temperature, usually 298 K, even though instantaneous

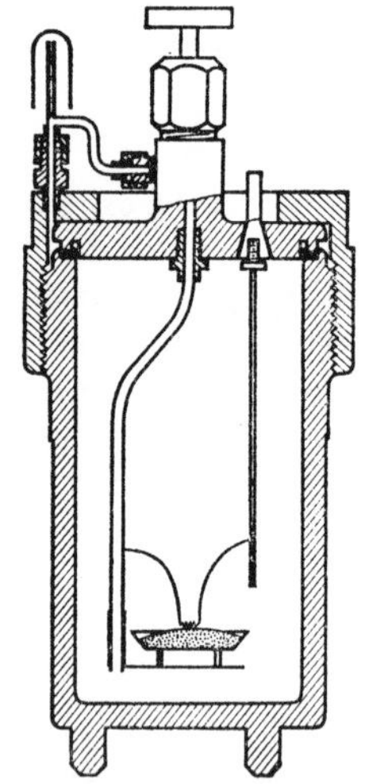

FIG. 3.1 A typical calorimeter bomb for use with pure oxygen. The sample dish, igniting wire, and oxygen port are clearly visible. Not shown is the insulated water bath in which the bomb is immersed. There are facilities for stirring and accurately measuring the temperature of the water. (Reproduced from Jessup, R. S., *J. Res. Nat. Bur. Standards*, **21**, 475 (1938).)

temperatures of the reacting compounds would be very much higher for short periods of time. Numerous corrections must be applied before an accurate value of ΔU may be calculated. This is usually then converted to a value for ΔH^0_{298}, the *standard enthalpy change* for the reaction, using the method of 2.2.1. For the more straightforward reactions, a precision of 1 in 10 000 is not only possible, but necessary. In many cases, however, combustion is incomplete, or slow, or otherwise difficult, and precision is much impaired.

Table 3.1 shows a selection of ΔH^0_{298} values for the combustion of a number of elements and organic compounds. (In the case of the elements, the values represent enthalpies of formation of the oxides.)

TABLE 3.1 *A selection of values of ΔH^0_{298} of combustion*

Material	State	Oxides	ΔH^0_{298} of combustion, kJ/mol
Al	c	Al_2O_3	−1675.6
Ca	c	CaO	−635.1
Cr	c	Cr_2O_3	−1141.0
Mo	c	MoO_3	−745.4
Re	c	Re_2O_7	−1238.0
Si	c	SiO_2	−910.0
C_6H_6	l	CO_2, H_2O	− 3273
C_2H_6	g	do	−1541.4
CH_4	g	do	−882.0
$(C_2H_5)_2O$	l	do	−2727

Example 3.1 Cobble, Smith and Boyd [1] performed the reaction

$$2\,Tc\,(c) + \tfrac{7}{2}\,O_2\,(g) + H_2O\,(l) + aq \longrightarrow 2\,HTcO_4\ \text{(dil. solution)}$$

in a bomb calorimeter.

In a typical experiment, 56.39 mg of technetium caused a temperature rise of 0.057 2°C on combustion. Electrical calibration of the calorimeter shows its 'water equivalent' to be 5847 J/K, so this experiment gives a value of

$$\Delta U = -5847\,(0.057\,2)2(\text{A.W.})/(56.39)10^{-3} = -1172\ \text{kJ/mol.}$$

(A.W. = atomic weight.) This leads to a value of $\Delta H = -1180$ kJ/mol (which should be checked by the student). Moreover, the heat of (dilute) solution of Tc_2O_7 to form dilute pertechnic acid is known to be $\Delta H = -51.0$ kJ/mol Tc_2O_7. Using this datum, it is then possible to calculate the standard heat of formation of Tc_2O_7 (c), which turns out to be $\Delta H^0_f(Tc_2O_7\,(c)) = -1129$ kJ/mol. (This too should be checked.)

Fluorine-bomb calorimetry is a fairly recent development. Until the early sixties, reliable enthalpy data concerning fluorides was very scarce, principally because elemental fluorine gas is so very reactive. Finally, W. N. Hubbard and co-workers overcame the many difficulties of calorimeter design and experimental technique. Enthalpies of formation have been determined from direct fluorination experiments. Now that interplanetary travel is being developed so rapidly, very high energy fuel systems are essential. Fluorine containing compounds show great promise, and of course, full thermochemical data are a prerequisite for any further development. Table 3.2 shows the enthalpies of formation of a selection of fluorides.

Example 3.2 Hubbard also found a value for ΔH^0_{298} of −885 kJ/mol for the reaction

I $BN\,(c) + \tfrac{3}{2}\,F_2\,(g) \longrightarrow BF_3\,(g) + \tfrac{1}{2}\,N_2\,(g)$.

Find $\Delta H^0_{f,298}$ for BN (c) from this information.

From the table we find that for

II $B\,(c) + \tfrac{3}{2}\,F_2\,(g) \longrightarrow BF_3\,(g)$, $\Delta H^0_{298} = -1136$ kJ/mol. The equation for formation of boron nitride,

$$B\,(c) + \tfrac{1}{2}\,N_2\,(g) \longrightarrow BN\,(c),$$

is the outcome of the recipe II–I, and so

$$\Delta H^0_{f,298}(BN(c)) = -1136 - (-885)$$
$$= -251 \text{ kJ/mol}.$$

Without an established fluorine technique, such a value would not be available. The reasons are not difficult to seek. Boron reacts very slowly with nitrogen, and the reaction is quite unsuitable for calorimetric measurement. Moreover, once formed, boron nitride has very good high temperature resistance to oxidation, and so fluorination is one of the few reactions suitable for study.

TABLE 3.2 *Enthalpies of formation of some fluorides*[a]

Compound	State	$\Delta H^0_{f,298}$ kJ/mol
AlF_3	c	−1499
AsF_5	g	−1237
BF_3	g	−1136
CdF_2	c	−700.4
HfF_4	c	−1930
MgF_2	c	−1124
MoF_6	l	−1558
RuF_5	c	−892.9
SiF_4	g	−1615
TaF_5	c	−1904
TiF_4	c	−1649
UF_6	c	−2186
ZnF_2	c	−764.4
ZrF_4	c	−1911

[a] Taken from *J. phys. Chem.* **70**, 7 (1966), and references contained therein.

3.2 Concepts of Heat Capacity

In many respects, water is an exceptional material. The specific heat of water is one exceptional property for which we all have cause to be thankful. The high specific heat of water is responsible, among other things, for the comfort of a hot-water bottle on a winter's night, and for the equable climate of oceanic regions. Questions of specific heat will concern us frequently. The property we shall use, rather than the specific heat, is the *molar heat capacity*, which is the energy required to raise the temperature of *one mole* of a substance by one degree Kelvin. Defined in terms of a differential, the heat capacity is

$$C = \frac{dq}{dT}, \text{ J/mol K}.$$

As it stands, this definition is not complete, because dq needs to be specified.

This dq depends on the actual process followed. If the volume is constant,

$$C_V = \frac{dq_V}{dT} = \frac{dU}{dT}.$$

At constant pressure,

$$C_P = \frac{dq_P}{dT} = \frac{dH}{dT}.$$

We recall that U and H are closely linked, and so it follows that C_V and C_P must also be linked. That is,

$$C_P = \frac{dH}{dT} = \frac{d(U + PV)}{dT} = \frac{dU}{dT} + \frac{d(PV)}{dT}.$$

For *a gas which behaves ideally*, $PV = RT$, and $d(PV) = R\,dT$.

Thus $$C_P = C_V + R. \qquad \text{(ideal gas) (3.1)}$$

The case of solids and liquids is not as simple. The equation of state is more complex, due to what is known as *internal pressure*, which arises from close-range interatomic forces. These are neglected in the case of ideal gases.

For gases, the heat-capacity increases both with the mass of each atom in the molecule, and with the degree of molecular complexity. Table 3.3 gives values of C_P for some substituted silanes.

TABLE 3.3 *Values of C_P for some silanes at 298* K[a]

Compound	C_P, J/mol K
SiH_4	42.8
SiH_3Cl	51.0
SiH_2Cl_2	62.0
$SiHCl_3$	75.5
$SiCl_4$	90.5
$Si(CH_3)Cl_3$	102.4
$Si(CH_3)_2Cl_2$	112.9
$Si(CH_3)_3Cl$	124.4
$Si(CH_3)_4$	138.6

[a] MIKAWA, Y. *Nippon Kagaku Zasshi* **81**, 1512 (1960).

The value for SiH_4 itself is fairly small. On substituting the hydrogen atoms by four chlorine atoms (A.W. = 35.5) the value of C_P roughly doubles. On the other hand, if four methyl groups (C + 3H = 15) are substituted, the increase is twice as great. Although the methyl groups are lighter than the chlorine, they are much more complex. Values for the intermediate compounds change evenly from one extreme to the other. Values of entropy, to be discussed in Chapter 5, are closely linked to heat capacity, and show similar qualitative variations.

Let us now suppose that we are asked to calculate the thermal energy required to heat a sample of methyl chloride (at constant pressure) over a range of temperature, say from T_1 to T_2. From the definition of C_P,

$$dq_P = C_P\,dT$$

and so

$$q_P = \int_{T_1}^{T_2} C_P\,dT. \tag{3.2}$$

There are three possible courses of action.

(a) *Temperature range is small*

In general, C_P varies with temperature, but for a small change, we can take an average value of C_P, and then

$$q_P = C_P\int_{T_1}^{T_2} dT = C_P(T_2 - T_1)\ \text{joule/mol}.$$

(b) *Values of C_P are known in tabular form*

The so-called JANAF tables [2] are an excellent compilation of thermodynamic data in tabular form. Table 3.4 is an extract of the heat capacity data for methyl chloride over a range of temperatures.

TABLE 3.4 *Heat capacity data for* CH_3Cl, *from* 300–1000 K

T, K	C_P, J/mol K
300	40.82
400	48.10
500	55.09
600	61.25
700	66.60
800	71.26
900	75.33
1000	78.90

By drawing a simple graph of C_P against T, and measuring the area, $q_P = \int C_P\,dT$ may easily be determined.

From Fig. 3.2 it is found that 41.76 kJ are needed to heat one mole of CH_3Cl from 350 K to 1000 K.

(c) *C_P is known as a power series*

As can be seen from Fig. 3.2, C_P varies in a non-linear fashion with temperature. The number of terms commonly used to describe C_P varies from two to four. High temperature reactions are becoming increasingly important, and as a result wider variations in C_P are encountered, so we shall use expressions of the form

$$C_P = a + bT + cT^2 + dT^3. \tag{3.3}$$

The terms a, b, c and d are constant for a particular material, and T is in K. For methyl chloride

$$C_P = 12.76 + (10.86)10^{-2}T - (5.205)10^{-5}T^2 + (9.623)10^{-9}T^3,$$

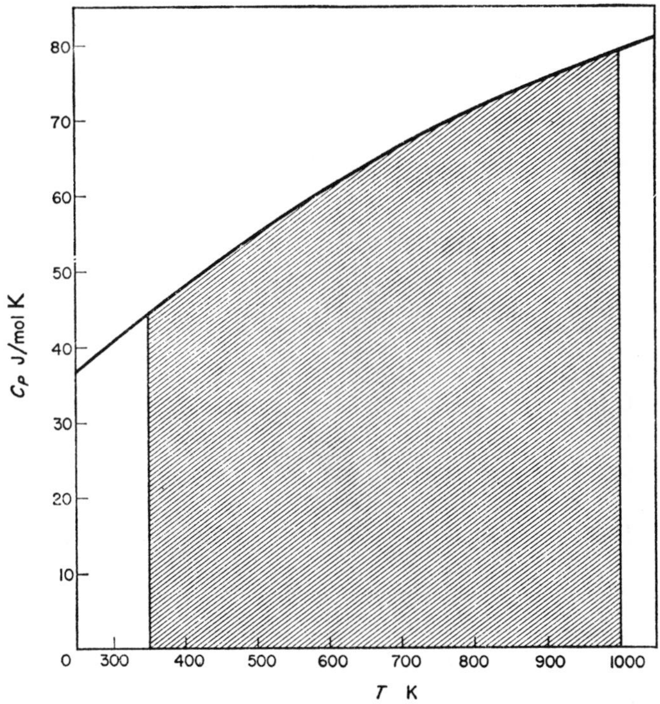

FIG. 3.2 Determination of thermal energy needed to raise 1 mole of CH_3Cl from 350 to 1000 K.

and this is valid over the range 273–1500 K. In this case, $a = 12.76$, and so on. Table 3.5 gives data for a selection of materials, in terms of eq. 3.3.

In order to calculate q_P, we must use eq. 3.2,

$$q_P = \int_{T_1}^{T_2} C_P \, dT.$$

Using eq. 3.3, this becomes

$$\begin{aligned} q_P &= \int_{T_1}^{T_2} (a + bT + cT^2 + dT^3) \, dT \\ &= \left[aT + bT^2/2 + cT^3/3 + dT^4/4 \right]_{T_1}^{T_2}. \end{aligned}$$

Using methyl chloride as an example, with data from Table 3.5, we have

$$\begin{aligned} q_P &= \Big[12.76T + (10.86)10^{-2}T^2/2 \\ &\qquad - (5.205)10^{-5}T^3/3 + (9.623)10^{-9}T^4/4 \Big]_{350}^{1000} \\ &= 8297 + 47\,650 - 16\,602 + 2368 \\ &= \mathbf{41.71\ kJ/mol.} \end{aligned}$$

This value agrees well with the previous graphical value of 41.76 kJ/mol.

TABLE 3.5 *Molar heat capacities for a selection of gases*[a] *according to the equation* $C_P = a + bT + cT^2 + dT^3$

Compound		a	10^2b	10^5c	10^9d	Valid, K
Ammonia	NH_3	27.55	2.563	0.9900	−6.686	273–1500
Benzene	C_6H_6	−36.19	48.44	−31.55	77.57	273–1500
Carbon monoxide	CO	27.11	0.655	−0.1000	—	273–3800
Carbon dioxide	CO_2	22.24	5.979	−3.498	7.464	273–1800
Ethane	C_2H_6	6.895	17.25	6.402	7.280	273–1500
Ethanol	C_2H_5OH	19.9	20.95	−10.372	20.04	273–1500
Ethylene	C_2H_4	3.95	15.63	−8.339	17.66	273–1500
Hydrogen	H_2	29.09	−0.1916	0.4000	−0.870	273–1800
Hydrogen fluoride	HF	30.13	−0.494	0.6594	−1.573	273–2000
Hydrogen chloride	HCl	30.31	−0.761	1.326	−4.335	273–1500
Methane	CH_4	19.87	5.021	1.268	−11.00	273–1500
Methyl chloride	CH_3Cl	12.76	10.86	−5.205	9.623	273–1500
Nitrogen	N_2	27.32	0.6226	−0.0950	—	273–3800
Oxygen	O_2	25.46	1.519	−0.7150	1.311	273–1800
Water	H_2O	32.22	0.1920	1.054	−3.594	273–1800

[a] Taken from KOBE, K. A. 'Thermochemistry for the petrochemical industry', *Petroleum Refiner* (November 1954). Units are in J/mol K.

3.2.1 COMBUSTION AND FLAME TEMPERATURES

Flames are exceedingly complex. As a result, it would seem that to attempt to calculate the temperature of, say, the combustion gases in a turbojet engine, would be quite useless. It is not impossible, however, as we shall see. Moreover, by making one or two intelligent simplifications, we shall be able to determine a *theoretical maximum flame temperature*, which, although it will need to be carefully qualified, will be a starting point for further analysis. But first, picture a high temperature flame. The premixed gases are completely reacted in a few

TABLE 3.6 *Maximum flame temperatures attainable by using various fuel gases*

Fuel/oxidant mixture	Maximum temperature, K
Propane–air	2200
Hydrogen–air	2320
Acetylene–oxygen	3430
Cyanogen–oxygen	4910

milliseconds, rapidly reaching a high temperature. Many chemical intermediates and radicals are present, and radiation of energy occurs immediately after the reaction is initiated. Some actual flame temperatures are given in Table 3.6, and serve to show the range of nearly 3000 deg which is commonly met.

To add to the apparent complications, we must remind ourselves that ΔH for a reaction changes slightly with temperature, and so also does C_P for the gaseous products, which absorb much of the available thermal energy. However, the first and principal simplification is to apply the First Law. If we assume an adiabatic process (no heat lost from the flame itself, either by conduction or radiation) then we can use the fact that H is a function of state. The overall process is

$$\text{Fuel} + \text{oxidant (298 K)} \longrightarrow \text{Products (several thousand degrees)}$$

but it may be performed in two (imaginary) steps.

I Fuel + oxidant (298 K) ⟶ Products (298 K) ($\Delta H = -$ve)

II Products (298 K) ⟶ Products (several thousand degrees) ($\Delta H = +$ve).

That is, *the thermal energy available from the exothermic reaction is then used to heat up the products* from 298 K to their final temperature. If we use eq. 3.2, which is

$$q_P = \int_{T_1}^{T_2} C_P \, \mathrm{d}T,$$

we have to solve for T_2, knowing q_P (= ΔH of the reaction), T_1 (= 298 K), and C_P (of the products). One last point should be made before we make a sample calculation. *The 'products' must be carefully specified.* Thus, of the common flame products, CO_2 dissociates extensively above 1000 K into CO and O_2, and water breaks up to give OH radicals, O, and H atoms above about 3500 K.

Example 3.3 In recent years, the simple 'flame test', whereby atoms of say sodium are excited in a flame, giving a characteristic yellow colour, has been developed as a sophisticated and sensitive instrumental technique (flame emission spectrophotometry). Sensitivity depends on dissociation of the injected materials into free atoms in order that the characteristic atomic emissions can be given. This in turn demands high flame temperatures. The combination of acetylene fuel with nitrous oxide as oxidant has proved highly successful for this purpose. What temperature is possible in theory? The best mixture would correspond to the equation

$$C_2H_2 + 3\,N_2O \longrightarrow 2\,CO + H_2O + 3\,N_2.$$

(If more N_2O were used, would CO_2 be formed?)

Using enthalpies of formation from Table 2.1, we have (using eq. 2.10)

$$\begin{aligned} \Delta H^0 &= 2(-110.5) + (-241.8) - (226.7) - 3(82.05) \\ &= -935.7 \text{ kJ}. \end{aligned}$$

Notice that the enthalpies of formation of both C_2H_2 and N_2O are positive, and energy is released on their breakdown. This energy is now available for heating up the gaseous products. (We shall assume that water is undissociated.)

Now $q_P = \int_{T_1}^{T_2} C_P \, \mathrm{d}T$ becomes in this case

$$+935\,700 = \int_{298}^{T_2} C_P(2\,CO + H_2O + 3\,N_2) \, \mathrm{d}T.$$

By determining C_P for each product gas, and adding, C_P for the products is found to be

$$\begin{aligned} C_P &= 2\,C_P(\mathrm{CO}) + C_P(\mathrm{H_2O}) + 3\,C_P(\mathrm{N_2}) \\ &= 165.3 + (4.628)10^{-2}T - (0.6870)10^{-5}T^2. \end{aligned}$$

Therefore

$$935\,700 = \int_{298}^{T} (165.3 + (4.628)10^{-2}T - (0.6870)10^{-5}T^2)\,\mathrm{d}T.$$

On integrating this expression, and inserting the limits, we have the cubic equation

$$(22.90)10^{-7}T^3 - (23.14)10^{-3}T^2 - 165.3T + 987\,100 = 0,$$

from which we find that $T = 4400$ K. The result is encouragingly high, and the nitrous oxide-acetylene flame is widely used. Temperatures of 3230 K are achieved in practice, and this shows that significant errors arise from neglecting water decomposition and radiation losses.

3.2.2 Variation of Reaction Enthalpies With Temperature

Suppose that a new vapour phase hydrogenation process is to be launched on a pilot plant scale. The reaction equation is known, and so ΔH^0_{298} is easily calculated from formation data. However, the reaction proceeds at a convenient rate at, say, 600 K. Will ΔH^0 change? By much? In which direction?

Heat capacities are clearly involved in the heating of reactants and cooling of products between 600 and 298 K. The simplest analysis is as follows. The change of ΔH with temperature is

$$\begin{aligned} \frac{\mathrm{d}(\Delta H)}{\mathrm{d}T} &= \frac{\mathrm{d}(H_{\text{products}} - H_{\text{reactants}})}{\mathrm{d}T} \\ &= \left(\frac{\mathrm{d}H}{\mathrm{d}T}\right)_{\text{products}} - \left(\frac{\mathrm{d}H}{\mathrm{d}T}\right)_{\text{reactants}} \\ &= C_P\,(\text{products}) - C_P\,(\text{reactants}). \end{aligned}$$

This simple result is often written

$$\boxed{\frac{\mathrm{d}(\Delta H)}{\mathrm{d}T} = \Delta C_P}. \qquad (3.4)$$

In an exactly analogous way, it can be shown that

$$\frac{\mathrm{d}(\Delta U)}{\mathrm{d}T} = \Delta C_V. \qquad (3.5)$$

These equations are often known as **Kirchhoff's Equations,** after their first protagonist. Equation 3.4 is more widely used than eq. 3.5, and our attention will be confined to it from now on.

Equation 3.4 may be integrated.

$$\int_{T_1}^{T_2} \mathrm{d}(\Delta H) = \int_{T_1}^{T_2} \Delta C_P\,\mathrm{d}T,$$

$$\boxed{\Delta H_2 - \Delta H_1 = \int_{T_1}^{T_2} \Delta C_P\,\mathrm{d}T}. \qquad (3.6)$$

Normally ΔC_P will take the form of a combination of C_P expressions in T, which must then be integrated. For small temperature ranges, mean values for C_P may be taken, and then

$$\Delta H_2 = \Delta H_1 + \Delta C_P(T_2 - T_1). \quad (3.7)$$

Example 3.4 Using a flow calorimeter, Lacher [3] has studied the hydrogenation of methyl chloride,

$$CH_3Cl + H_2 \rightarrow CH_4 + HCl.$$

He found that, at 248°C (521 K), $\Delta H^0_{521} = -82.32$ kJ/mol. Before this result can be incorporated into tables of standard data, adjustment to the standard temperature (298 K) must be made. For this calculation we will use values of C_P at 400 K, of

$$\begin{aligned} C_P(CH_4\,(g)) &= 41.28 \text{ J/mol K}, \\ (HCl\,(g)) &= 29.11 \text{ J/mol K}, \\ (CH_3Cl\,(g)) &= 48.46 \text{ J/mol K}, \\ (H_2\,(g)) &= 28.56 \text{ J/mol K}. \end{aligned}$$

Use eq. 3.7, with '1' denoting the experiment at 521 K, and '2' denoting the process at 298 K.

Then

$$\begin{aligned} \Delta C_P &= C_P \text{ (products)} - C_P \text{ (reactants)} \\ &= 41.28 + 29.11 - 48.46 - 28.56 \\ &= -6.63 \text{ J/mol K}, \end{aligned}$$

and so

$$\begin{aligned} \Delta H^0_{298} &= \Delta H^0_{521} + (-6.63)(298 - 521)/1000 \\ &= -82.32 + 1.477 \\ &= \mathbf{-80.84 \text{ kJ/mol}}. \end{aligned}$$

Thus we see that the reaction would be nearly 2% less exothermic if it were to be carried out at 298 K. As a general guide, although the heat capacities tend to be self-cancelling, there are significant changes in ΔH which must be accounted for, especially when considering reactions in solution.

3.3 Bond Energies

A chemical reaction often involves the breaking of some bonds, and the making of others. It seems feasible, therefore, that there should be a close link between the stability of a bond and the enthalpy change involved in breaking it. We can represent the breaking of the X—Y bond in one mole of the molecule XY (not necessarily diatomic) by the equation

$$XY\,(g) \rightarrow X\,(g) + Y\,(g), \quad \text{enthalpy change} = \Delta H^0_{298}.$$

If the fragments and the compound are gases in their standard states, at 1 atm and 25°C, then we can define the **standard bond dissociation energy** DH^0 for the bond X—Y as

$$DH^0(X\text{—}Y) = \Delta H^0_{298}.$$

For a particular bond, the value of DH^0 will depend upon the local environment, upon the presence or absence of nearby electrophilic groups, and so on. For example, the two O—H bonds of water can be broken in succession.

$$HO\text{—}H\ (g) \rightarrow HO\ (g) + H\ (g),\quad \Delta H = +494\ kJ/mol,$$
$$O\text{—}H\ (g) \rightarrow O\ (g) + H\ (g),\quad \Delta H = +431\ kJ/mol.$$

The standard bond dissociation energy (BDE) in each case is therefore DH^0 (HO—H) = 494 kJ, and DH^0(O—H) = 431 kJ. Methane provides a second, more extreme example. The successive removal of hydrogen atoms may be summarized:

$$DH^0(CH_3\text{—}H) = 435\ kJ,$$
$$DH^0(CH_2\text{—}H) = 444\ kJ,$$
$$DH^0(CH\text{—}H) = 444\ kJ,$$
$$DH^0(C\text{—}H) = 339\ kJ.$$

It is perhaps surprising not that the values vary, but that they are as close as they are. Values of DH^0 for a small number of specific bonds have been determined with difficulty using such techniques as mass spectrometry, and ultraviolet and visible spectroscopy. On the whole, however, accurate values are rather few and far between, and as there are far more bonds than there are chemical compounds, the prospect for complete tables of DH^0 is slight indeed.

3.3.1 Average Bond Dissociation Energies

An alternative approach is to consider the complete atomization of a compound, such as methane, into its *constituent atoms in the gas phase.*

$$CH_4\ (g) \rightarrow C\ (g) + 4\ H\ (g).$$

This reaction involves the breaking of all four C—H bonds, and so the enthalpy change will be a sum of all four standard bond dissociation energies. However, we can write, from eq. 2.10,

$$\Delta H^0 = \Delta H_f^0(C\ (g)) + 4\ \Delta H_f^0(H\ (g)) - \Delta H_f^0(CH_4\ (g)). \tag{3.8}$$

Enthalpies of atomization, or formation of atoms, are much easier to determine than specific bond energies, and are on the whole more accurate.

Values for several important elements are included in Table 2.1. On substituting values into eq. 3.8,

$$\Delta H^0 = 715.0 + 4(217.9) - (-74.85)$$
$$= 1661\ kJ.$$

The **average** bond dissociation energy, DH_{av}, can now be calculated. The above figure represents the breaking of *four* C—H bonds,

$$DH_{av}(C\text{—}H) = \Delta H^0/4 = 1661/4 = 415.3\ kJ/mol.$$

Let us now assume that for ethane, the six C—H bonds have an identical average bond dissociation energy. Atomization of ethane involves rupture of six C—H bonds, and one C—C bond. Now, ΔH^0 for ethane atomization, which can be calculated as before, is

$$\Delta H^0 = 2\ \Delta H_f^0(C\ (g)) + 6\ \Delta H_f^0(H\ (g)) - \Delta H_f^0(C_2H_6\ (g))$$

which also $= 6\ DH_{av}(C\text{—}H) + DH_{av}(C\text{—}C),$

and so a value for the C—C bond may be calculated. In this way, successive pairs of hydrocarbons yield a series of values of DH_{av} for the C—H and C—C bonds. These are summarized in Table 3.7.

It is clear that such bonds are of equal strength in whatever molecule they occur.

TABLE 3.7 *Values of DH_{av}(C—H) and DH_{av}(C—C) calculated from normal alkanes, in kJ/mol*

n-Alkane	DH_{av}(C—H)	DH_{av}(C—C)
CH_4	415.5	—
CH_4, C_2H_6	415.5	329.7
C_2H_6, C_3H_8	413.0	343.9
C_3H_8, C_4H_{10}	411.3	350.2
C_4H_{10}, C_5H_{12}	413.0	345.2
C_5H_{12}, C_6H_{14}	412.5	346.4

Although 'irregular' environments will cause serious discrepancies, it is none the less possible to use such average values to predict properties of compounds of a given structure. Such average bond energies have been determined for many other bonds, and they provide a rough and ready guide to heats of formation, and of combustion, if these are otherwise unknown. Table 3.8 summarizes such values.

TABLE 3.8 *Average bond dissociation energies for a selection of single and multiple bonds*

	I	Br	Cl	F	O	C	H
H	297	368	431	569	460	414	435
C	226	289	343	444	339	343	
O	—	—	205	184	146		
F	243	251	255	159			
Cl	209	218	243		kJ/mol		
Br	180	192					
I	151						

C═C	611 kJ/mol
C≡C	833 kJ/mol
C═O	741 kJ/mol
C≡N	879 kJ/mol

Example 3.5 Ethyl hydroperoxide C_2H_5OOH (g) is rumoured to have a value of $\Delta H_f^0 = -167.4$ kJ/mol. Is this feasible?

First, let us write down the atomization process, as

$$C_2H_5OOH\,(g) \rightarrow 2\,C\,(g) + 6\,H\,(g) + 2\,O\,(g).$$

This involves breaking of five C—H, one C—C, one C—O, one O—O, and one O—H bond, and so, from Table 3.8,

$$\Delta H^0 = 5(414) + 343 + 339 + 146 + 460 = 3358 \text{ kJ/mol}.$$

In addition, we can use enthalpies of formation, and

$$\begin{aligned}\Delta H^0 &= 2\,\Delta H_f^0(\text{C (g)}) + 6\,\Delta H_f^0(\text{H (g)}) + 2\,\Delta H_f^0(\text{O (g)}) \\ &\quad -\Delta H_f^0(\text{C}_2\text{H}_5\text{OOH (g)}), \\ &= 2\,(715.0) + 6\,(218.0) + 2\,(249.4) - \Delta H_f^0(\text{C}_2\text{H}_5\text{OOH}).\end{aligned}$$

By eliminating ΔH^0 we find the value

$$\Delta H_f^0(\text{C}_2\text{H}_5\text{OOH}) = 3237 - 3358 = \mathbf{-121\ kJ/mol},$$

which is reasonable, considering that it is a small difference between two large figures.

Example 3.6 Vaughn and Muetterties [4] found that $\Delta H = -718.4$ kJ/mol for the formation reaction

$$\text{S (c)} + 2\ \text{F}_2\ \text{(g)} \longrightarrow \text{SF}_4\ \text{(g)}.$$

What is the average bond dissociation energy of the S—F bond in SF_4?

Complete bond dissociation implies the reaction

$$\text{SF}_4\ \text{(g)} \longrightarrow \text{S (g)} + 4\ \text{F(g)},$$

for which

$$\begin{aligned}4\,DH_{av}(\text{S—F}) &= \Delta H^0 \\ &= \Delta H_f^0(\text{S (g)}) + 4\,\Delta H_f^0(\text{F (g)}) - \Delta H_f^0(\text{SF}_4\ \text{(g)}) \\ &= 274.9 + 4(79.1) - (-718.4) \\ &= 1310, \\ \therefore\ DH_{av}(\text{S—F}) &= 1310/4 = \mathbf{327.5\ kJ/mol}.\end{aligned}$$

REFERENCES

[1] Cobble, J. W., Smith, Wm. T., Jr. and Boyd, G. E. *J. Am. chem. Soc.* **75,** 5777 (1953).
[2] *JANAF Thermochemical Tables,* Publication PB 168–370, Clearinghouse for Federal Scientific and Technical Information, Springfield, Virginia.
[3] Lacher, J. R., Emery, E., Bohmfalk, E. and Park, J. D. *J. phys. Chem.* **60,** 492 (1956).
[4] Vaughn, J. D. and Muetterties, E. L. *J. phys. Chem.* **64,** 1787 (1960).

PROBLEMS

3.1 A sample of 39.44 mg of liquid phenyl acetylene, $C_6H_5.CCH$ (MW = 102.14) is burned in pure oxygen in a bomb calorimeter at 25°C. Thermal energy is released, equivalent to 1656 J of electrical energy. What is, per mole, q_V, w, ΔU? Assuming that the products are CO_2 (g) and H_2O (l), calculate ΔH.

3.2 A flow calorimeter is operated at 1 atm and 248°C. In the formation of HCl gas from H_2 and Cl_2, a typical run showed that $(3.913)10^{-4}$ mole HCl/min liberated 36.56 J/min of energy. What is ΔH_f^0(HCl) at 248°C? If average values of C_P are

$$\begin{aligned} &HCl\ (g): 29.1\ J/mol\ K,\\ &H_2\ \ (g): 28.8\ J/mol\ K,\\ &Cl_2\ \ (g): 33.9\ J/mol\ K, \end{aligned}$$

determine $\Delta H^0_{f,298}(HCl)$.

3.3 The Sachse process is used for the production of acetylene from natural gas:

$$2\,CH_4 \overset{1500°C}{\rightleftharpoons} C_2H_2 + 3\,H_2.$$

From Table 2.1, ΔH^0_{298} is found to be $+376.4$ kJ/mol C_2H_2. Using the following average values of C_P, estimate ΔH^0 at the operating temperature.

$$\begin{aligned} C_P(CH_4) &= 61.6\ J/mol\ K,\\ (C_2H_2) &= 62.7\ J/mol\ K,\\ (H_2) &= 31.2\ J/mol\ K. \end{aligned}$$

3.4 The use of average heat capacities in the previous problem leads to poor accuracy, especially over the very large temperature range involved. Using $C_P(T)$ functions from Table 3.5, make a better estimate of the actual ΔH^0 value at 1500 °C.

3.5 Dungeness 'B' nuclear station is centred around the first large 'advanced gas-cooled reactor' (AGR). The primary coolant for the uranium oxide fuel elements is high-pressure CO_2 gas. It enters the core at 561 K and leaves at 918 K. It is obviously essential to know the amount of CO_2 needed to remove 1200 *megawatts* from the core. Use C_P data to find (a) $(H_{918}—H_{561})$ in kJ/mol, algebraically or graphically, and (b) how many kg of CO_2 must pass per second.

3.6 Mercury boils at 629.7 K, under atmospheric pressure, and $\Delta H_{vap} = 59.27$ kJ/mol at that temperature. A mercury distillation plant runs at 673.0 K; what is ΔH_{vap} at that temperature? Average values of C_P for liquid and vapour are 27.4 and 20.6 J/mol K, respectively.

3.7 Ammonia fuel with pure fluorine oxidant gives a very high-temperature flame:

$$NH_3 + \tfrac{3}{2}\,F_2 \rightarrow \tfrac{1}{2}\,N_2 + 3\,HF.$$

Assuming that the molar heat capacities of the products may be averaged over the temperature range to

$$\begin{aligned} C_P(N_2) &= 32.55\ J/mol\ K,\\ C_P(HF) &= 31.60\ J/mol\ K, \end{aligned}$$

calculate the theoretical flame temperature starting at 25 °C. Neglect the effects of radiation losses and product dissociation.

3.8 For ethylene burning with pure oxygen to give CO and H_2O, the theoretical flame temperature approaches 5000 K. Make an accurate calculation of this for reactants at 25 °C, using the C_P equation in Table 3.5 for CO, and the following one for H_2O:

$$C_P(H_2O) = 29.16 + (1.449)10^{-2}T - (.2022)10^{-5}T^2.$$

(This will involve solution of a cubic equation.)

3.9 Compare the value of -131.6 kJ/mol for the standard enthalpy change of formation of gaseous 2-methyl propane (C_4H_{10}) with that calculated from bond energy data (Tables 3.8 and 2.1).

3.10 The standard enthalpy change for formation of KrF_2 (g) is +57.3 kJ/mol. Using atomization data from Table 2.1, determine the bond dissociation energy for the Kr—F bond.

4

Spontaneous Changes

So far, we have learnt to follow, by means of the First Law, the transfer of energy, and its conversion from one form to another. A question that is often asked is 'will this particular process tend to occur or not?' The First Law *provides no answer to this kind of question*, and we must look further for a guide to reactivity. By considering various *spontaneous processes*, we shall find that there are two driving forces in nature. The drive towards minimization of energy is one such directing influence, but there is also a tendency for materials to become *more physically disorganized.* The concept of entropy will be treated in more detail in Chapter 5, but is introduced here as a measure of the 'randomness' of materials; the tendency for entropy to increase is nature's second driving force.

4.1 Everyday Processes

Our daily lives consist of a multitude of more or less familiar occurrences, most of which are so common that we do not stop to consider the direction of these changes. Water runs downhill; a plate, dropped onto the floor, breaks and the pieces scatter; petroleum vapour and oxygen, when reminded by a small spark of their chemical obligations, react in a car engine to produce the usual mixture of exhaust gases; the exhaust gases so formed then disperse into the atmosphere.

Making snowballs will always make one's hands colder, because thermal energy always flows in a 'hot-to-cold' direction. No material can spontaneously gain thermal energy from colder surroundings. We take this for granted, but this simple fact is at the heart of the nature of entropy.

We must ask ourselves why the opposite processes do not occur. Can water flow, *of itself*, upwards? If we wait long enough, will the pieces of broken plate *spontaneously* reorganize themselves into their previous arrangement? Can carbon monoxide, carbon dioxide, and water vapour ever congregate, and react to form petroleum and oxygen? Has anything *ever* been known to get hot in the refrigerator? Of course these are 'foolish' questions, because *no one has yet seen these things happen.* In many instances, energy is converted from one form to another, the total energy being conserved in each case (First Law). Falling water represents the conversion of gravitational energy to kinetic, and then to thermal energy. (Water gains about 0.1 °C in temperature when it falls 40 m.) Chemical energy is converted to thermal energy in the cylinder of the petrol engine, although little if any energy is involved in the dispersion of cooled exhaust gases. In no case is the First Law of Thermodynamics *able to say anything about the direction* of these common occurrences.

4.2 Exothermicity, a Possible Criterion

The problem of spontaneity, as outlined above, occupied many eminent scientists in the latter half of the nineteenth century. In the case of chemical reactions carried out under the usual conditions of constant temperature and pressure, it was initially assumed that *if, and only if*, a reaction gave out heat, it occurred spontaneously. It turns out that a second factor is involved, and that both factors must be considered. By considering a number of examples in which ΔH is negative, or zero, or positive, this second factor, which is best described as a 'tendency to randomness', will become more familiar.

4.2.1 Spontaneous Exothermic Processes

Most exothermic chemical reactions, including oxygen combustions, are spontaneous in the sense that the reactants are unstable compared with the products. The reaction may well need a catalyst in order to make it occur at an appreciable rate, or perhaps a spark may be necessary to initiate the process. The rate may be very low, but the direction of the process is fixed.

Another example of a spontaneous, exothermic process will be familiar to skiers. Fresh snow is powdery, and affords smooth and easy skiing, but over a period of weeks, the conditions deteriorate. By a process of progressive cannibalization, large crystals grow at the expense of the small. For a given mass of snow, the surface area, and hence the surface energy, is greater for small crystals, and so their vapour pressure is very slightly higher. The overall effect is to convert the potential surface energy to thermal energy, by altering the size distribution of the ice crystals.

On the basis of such arguments as these, M. Berthelot and J. Thomson became convinced that exothermicity was the only factor involved in determining 'chemical affinity'. We shall see that many spontaneous processes have zero, and even positive values of ΔH, which clearly shows that this is only part of the story.

4.2.2 Spontaneous Processes Involving no Heat Change

(a) Suppose that the water in a covered swimming pool is allowed to come to complete rest, with no eddies or convection currents. If a drop of strong aqueous dye solution is carefully added, not dropped, at the surface, the colour will quickly spread. Similarly if a girl wearing scent walks quietly into a large room, it

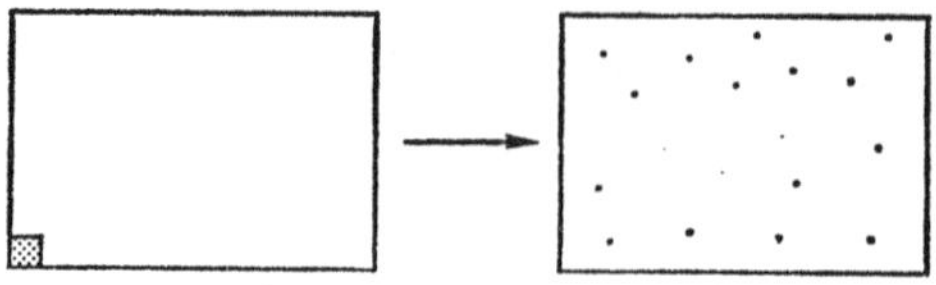

FIG. 4.1 Process in which concentrated matter disperses.

is only a matter of minutes before the molecules of scent will have diffused throughout the room, *even if the air is perfectly still.* (Such behaviour of gases and liquids is no different for near-ideal gases or solutions. It does not depend on non-ideality.) Both of these processes may be represented schematically, as shown in Fig. 4.1. Vanishingly small amounts of thermal energy are involved in such processes. Even as ideal behaviour is approached, and $\Delta H = 0$, no diminution of 'reactivity' is observed.

(b) In 1843, Joule performed a classic experiment which is depicted in Fig. 4.2. Dry air was introduced into the left-hand container, whereas the right-hand one was evacuated. The temperature was read carefully and noted, and then tap C opened. The gas immediately rushed through the stopcock and filled all the available space. Joule noticed no change of temperature of the water, even with

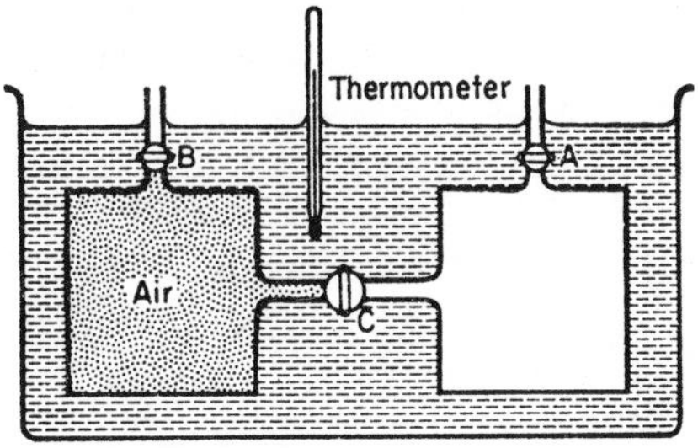

FIG. 4.2 Joule's gas expansion experiment. On allowing gas to flow into the right-hand container, no temperature change was observed.

his 'delicate thermometer'. (Joule's calorimeter was actually not sensitive enough to measure the very small temperature change involved, which is due solely to the non-ideality of the gas.) Helium, which is nearly ideal under ambient conditions, behaves in exactly the same way, although there is only a minute thermal change, and it is a reasonable assumption that a truly ideal gas would expand in a similar manner. Moreover, although the gas is turbulent when the tap is opened, no *net* work is done on the surroundings. This process is a natural, irreversible process, and occurs even though $q = w = 0$. Thus $\Delta U = 0$, and so the change of U with volume, under isothermal conditions, is zero. That is,

$$\left(\frac{\mathrm{d}U}{\mathrm{d}V}\right)_T = 0.$$

The important point to note, however, is that the *process occurs, although no heat is liberated.* (What general feature is common to processes (a) and (b)? The diagrams help.)

(c) A chemical reaction with $\Delta H = 0$ will now be discussed. Silver chlorite, $AgClO_2$, is unstable with respect to its elements.

$$AgClO_2\,(c) \longrightarrow Ag\,(c) + \tfrac{1}{2}\,Cl_2\,(g) + O_2\,(g).$$

It decomposes when heated gently, and shows no tendency to reform. However, it so happens that, quite by chance, ΔH is as near zero as can be measured. *Something other than exothermicity is the driving force behind this reaction.* It is significant that one and a half moles of gas are formed per mole of decomposing solid silver chlorite.

4.2.3 ENDOTHERMIC PROCESSES

(a) If a little diethyl ether is placed in an open dish, spontaneous evaporation occurs, and this requires thermal energy to maintain a constant temperature.

A 'cooling effect' is observed. Use is made of this effect when ethyl bromide is used as a local anaesthetic. Its boiling point, 38.4°C, is very close to normal body temperature, and so evaporation is very fast. The process is

$$C_2H_5Br\ (l) = C_2H_5Br\ (\mathbf{g}), \quad \text{and} \quad \Delta H^0 = 27.36\ \text{kJ/mol}.$$

Evaporation occurs to some extent from any liquid surface. Moreover, all evaporation processes are endothermic. In common with the silver chlorite decomposition mentioned above, *this process leads to a significant volume increase.*

(b) There are many spontaneous but endothermic chemical processes although few occur at laboratory temperatures. Consider one only.

Thionyl chloride is a very useful drying agent, because the products are completely gaseous, and contamination is usually negligible. It reacts with free water (and also with water of crystallization), according to the equation,

$$SOCl_2\ (l) + H_2O\ (l) = SO_2\ (\mathbf{g}) + 2HCl\ (\mathbf{g}).$$

The enthalpy change for the reaction is $\Delta H^0 = +50.02$ kJ/mol, and this reaction causes significant cooling under normal laboratory conditions. In this and similar *spontaneous endothermic* chemical reactions, *gaseous products arise from condensed reactants*. We shall see later that this 'gasification' is closely related to the increase of entropy or randomness associated with the reaction. *It seems that when an unfavourable, positive value of ΔH is overcome in a reaction, this is associated with randomization.*

4.3 The Second Driving Force

We have tested the idea that exothermicity is alone responsible for spontaneous reactions, and found it severely lacking. It does however contain some virtue, in that *sufficiently large* endothermicities do indicate stability of reactants. Thus thermal decomposition of Al_2O_3, or CaO does not occur even at quite high temperatures, and the decomposition reactions have very large positive ΔH^0 values (+1110 and +1270 kJ/mol of O_2, respectively).

That exothermicity is not the only factor is shown clearly in Section 4.2.2. Here, where $\Delta H = 0$, the changes were characterized by a spreading of at least some of the materials into a larger volume. Diffusion of the dye in the swimming-pool involved randomization, in a simple geometric sense. At the start, we at least knew where it was, with an uncertainty of only a millimetre or so. The decomposing silver chlorite started as a crystalline solid, but the products included one and a half moles of gas for every mole of solid silver, and this too represents an increase of randomness.

The concept of entropy is a difficult one. An understanding of it comes with repeated encounters in different contexts. Although a formal definition will be given later, it is sufficient for the time being to associate it pictorially with the *degree of randomness* of a material, as exemplified in the processes discussed above. It is possible to assign absolute numerical values to this entropy, S^0, which refer to individual materials in a particular physical state. In Table 4.1 values are given for a few compounds.

As a general guide, it will be found that entropy values increase during the transitions solid to liquid to gas, as demonstrated by the values for solid zinc, and for gaseous krypton, at one atmosphere. On the other hand, entropy will also increase with molecular complexity, as shown by $ZnSO_4.H_2O$ (c), which has an entropy

TABLE 4.1 *Numerical values of entropy for some compounds at* 25 °C

Material	Physical state	Entropy, S^0, J/mol K
C	crystal, graphite	5.73
ZnO	crystal	43.9
Zn	crystal	41.6
Kr	gas	164.0
$ZnSO_4.H_2O$	crystal	146.0
CO	gas	197.6

almost as great as that of gaseous krypton. The qualitative effects are none the less clear.

The descriptive approach to entropy adopted here is adequate for the *isothermal* processes chosen as examples. It is not easy, for example, to visualize the change of entropy which occurs when two bodies at differing temperatures come into thermal contact. (Entropy increases.) The mathematical approach to be developed in the next chapter will however cope with this and other difficulties which arise.

PROBLEMS

4.1 The Sachse process (see Problem 3.3) is highly endothermic. How can it ever be made to operate effectively? Use entropy values from Table 2.1 to develop your argument.

4.2 The example in 4.2.2(c), the decomposition of silver chlorite, proceeds with entropy increase. Using values of absolute molar entropies (S^0_{298}) from Table 2.1 for the various substances, calculate ΔS^0 for the process as written.

5

Entropy

M. JOURDAIN: 'What? When I say "Nicole, bring me my slippers and give me my nightcap", that's prose?'
LE MAÎTRE DE PHILOSOPHIE: 'Yes, sir.'
M. JOURDAIN: 'Gracious me! I've been talking prose for the last forty years and have never known it.'

(Molière, *Le Bourgeois Gentilhomme*)

Changes of entropy are occurring all around us, but for most of us, like M. Jourdain with his prose, awareness can come quite late in life. Although we cannot detect these changes with our physical senses, it is possible to calculate them from simple measurements. After defining entropy, and performing such calculations, the concept will become progressively more familiar.

5.1 Measurement of Entropy

5.1.1 THE SECOND LAW OF THERMODYNAMICS

The concept of entropy has come about slowly, with the accumulation of much human experience and is embodied in the so-called 'second law', *which is a statement of this practical experience, and no more.* It should be stressed that exceptions to the 'law' may yet be found. It may be stated in many different ways. Its origins are closely bound up with the science of heat engines.

There exists a thermodynamic function of state, called entropy, S. The entropy of system and surroundings together increases during all natural or irreversible processes. For a reversible process, the total entropy is unchanged.

Notice that entropy is not conserved; this is difficult to accept, although the evidence is in fact quite convincing.

A small change of entropy $\mathrm{d}S$, is measurable in terms of the thermal energy q_{rev} added to a system during a *reversible* process, and the (constant) temperature T, at which it is performed:

$$\boxed{\mathrm{d}S = \frac{\mathrm{d}q_{rev}}{T}}. \tag{5.1}$$

Because S is a function of state, the term $\mathrm{d}q$ in eq. 5.1 must be quite unambiguous. In fact, as we shall see, reversibility necessarily implies that maximum work is involved, and that q is specified.

5.1.2 Reversibility and Entropy

In general, energy or 'work' is the product of an *intensive factor* and an *extensive factor*.

Mechanical energy = force × distance,
Work of compression = pressure × volume change, and
Electrical energy = voltage × quantity of charge.

In each case, if the intensive factor (force, pressure or voltage) is opposed by a force, pressure or voltage which is *virtually equal* to it, then energy is transferred *reversibly*. The process will occur infinitely slowly, but will result in the transfer of maximum energy (maximum work). The eventual winners of a tug-of-war between two evenly matched sides will be working 'flat-out' for their win. The work done by the winning team depends on *the resistance they meet*, not on their own inherent strength. The maximum work will be done when they are virtually equal to their opponents, but the performance will then take a long time. They are working reversibly because a minute increase of effort by their opponents will *reverse the direction* of the change.

It is this concept of reversibility which must be applied when entropy changes are calculated. There is, however, no reason why the *actual* process should occur reversibly, because entropy is a function of state, and ΔS is the same, whatever the route. It is frequently necessary to devise, on paper, a reversible route merely to perform the entropy-change calculation. Let us apply these ideas to specific examples.

(a) *Isothermal expansion of gases*

The gas-expansion experiment of Joule, shown in Fig. 4.2, took place completely irreversibly; no work was done, because no resistance was offered from outside. The overall change is

$$\text{Gas}\ (T, V_1) \longrightarrow \text{Gas}\ (T, V_2).$$

Imagine a reversible path for this change in which an opposing piston offers virtually equal, but opposite pressure. As infinitesimal expansion occurs, so the opposing pressure drops by an infinitesimal amount, and expansion continues, albeit infinitely slowly. Negative work is being done on the gaseous system, by the piston (surroundings), and in order to maintain (constant) temperature heat must be added. Internal energy is then kept constant, and so for a small change,

$$\mathrm{d}U = 0 = \mathrm{d}q_{\text{rev}} + w = \mathrm{d}q_{\text{rev}} - P\,\mathrm{d}V$$

and so $\mathrm{d}q_{\text{rev}} = P\,\mathrm{d}V$.
On integrating, and making the ideal gas assumption,

$$\int \mathrm{d}q_{\text{rev}} = \int P\,\mathrm{d}V = nRT\int \frac{\mathrm{d}V}{V}$$

$$\therefore\ q_{\text{rev}} = nRT \ln V_2/V_1. \qquad (T\ \text{constant})\ (5.2)$$

From eq. 5.1, for a finite change,

$$\Delta S = \frac{q_{\text{rev}}}{T} = nR \ln V_2/V_1. \qquad (T\ \text{constant})\ (5.3)$$

The entropy change for the *irreversible* Joule experiment may thus be calculated *from the hypothetical, reversible process*, because S is a function of state only.

(b) *Reversible transfer of heat*

Consider a counter current heat exchanger, such as is used on many industrial plants. In essence this is a pair of pipes in close thermal contact (Fig. 5.1).

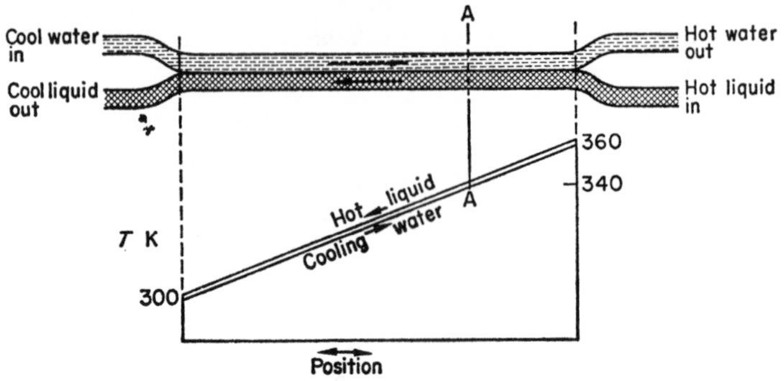

FIG. 5.1 Simplified heat exchanger. Reversible operation requires infinitely slow liquid movement; the outgoing liquid then has the same temperature as the inlet water.

If it were to be operated reversibly, the water and initially-hot liquid would be in continuous equilibrium, with at each point only an infinitesimal thermal gradient between the liquids. In such a case, the liquids are virtually stationary, and maximum energy will be transferred. Entropy changes at point A are

$$\mathrm{d}S_{\text{liquid}} = -\frac{\mathrm{d}q}{T_{\text{liquid}}},\ \mathrm{d}S_{\text{water}} = +\frac{\mathrm{d}q}{T_{\text{water}}},$$

and the total entropy change for liquid and water ('system and surroundings') is zero;

$$\mathrm{d}S_{\text{total}} = \mathrm{d}q\left(-\frac{1}{T_{\text{liquid}}} + \frac{1}{T_{\text{water}}}\right) = 0. \tag{5.4}$$

On the other hand, suppose that the liquids move at a finite speed. Now, T_{liquid} is greater than T_{water}, and eq. 5.4 gives $\mathrm{d}S_{\text{total}} > 0$. *Entropy has increased because, and to the extent that, the process is irreversible.*

(c) *An irreversible change*

Following heat treatment, a steel bar at 1000 K is quenched in warm water (333 K). This is a typical irreversible process. Consider the transfer of the first joule, which leaves the steel at 1000 K, and arrives in the water at 333 K. For purposes of calculation, this process could be considered in two reversible parts. First a joule is removed from the steel by contact with a thermostat set at $(1000 - \mathrm{d}T)$ K; then a joule is added to the water from a bath at $(333 + \mathrm{d}T)$ K. The entropy changes are, for the reversible process,

$$\Delta S_{\text{steel}} = -\frac{1}{1000},\ \Delta S_{\text{hot bath}} = +\frac{1}{1000},$$

$$\Delta S_{\text{water}} = +\frac{1}{333},\ \Delta S_{\text{cool bath}} = -\frac{1}{333},$$

$$\Delta S_{\text{total}} = 0. \qquad \text{(Reversible)}$$

In the irreversible quenching, only steel and water are involved, for which the entropy-changes are already calculated, so here

$$\Delta S_{\text{total}} = -\frac{1}{1000} + \frac{1}{333} = \mathbf{+\ 0.002\ J/\ K.} \qquad \text{(Irreversible)}$$

Once again, entropy has increased during the *irreversible* process.

5.1.3 Changes of Entropy with Temperature

So far, we have considered only isothermal processes. How do we calculate the entropy change for a gas which is heated? Now $\mathrm{d}S = \mathrm{d}q_{\text{rev}}/T$, and $C_P = \mathrm{d}q/\mathrm{d}T$, therefore as $\mathrm{d}q$ and $\mathrm{d}q_{\text{rev}}$ are identical,

$$\mathrm{d}S = \frac{C_P}{T}\,\mathrm{d}T$$

and

$$\Delta S = S_2 - S_1 = \int \frac{C_P}{T}\,\mathrm{d}T. \qquad (P \text{ constant}) \quad (5.5)$$

Over the temperature range where C_P may be considered constant,

$$S_2 - S_1 = C_P \int \frac{1}{T}\,\mathrm{d}T = C_P \ln T_2/T_1. \qquad (5.6)$$

In general, however, it is necessary to plot C_P/T against temperature. Equation 5.5 then tells us that the area under the curve between the two temperatures of interest gives the change of entropy.

Example 5.1 The average value of C_P for silane, SiH_4, between 340 and 360 K is 47.03 J/mol K. By how much does the entropy of four moles of silane increase when it is heated over this temperature range?

Using eq. 5.6,

$$\begin{aligned} S_{360} - S_{340} &= 4(47.03) \ln 360/340 \\ &= \mathbf{+1.08\ J/\ K.} \end{aligned}$$

For processes carried out at constant volume, an exactly analogous derivation using C_V may be followed to give

$$\Delta S = S_2 - S_1 = C_V \ln T_2/T_1. \qquad (V \text{ constant})$$

Again, a mean value of C_V is used.

5.1.4 An Adiabatic Compression

If a gas is effectively insulated during reversible compression, $q_{\text{rev}} = 0$, and so $\Delta S = 0$. This means that $\Delta U = w$, and the temperature rises. Does this not mean that the entropy also increases? The answer is no, because the effect of the temperature *increase* is exactly matched by the *reduction* in volume which accompanies it.

5.2 Absolute Entropies

So far, we have been concerned solely with *changes* of entropy. Is it possible to assign a zero entropy to a particular sample or state? In fact, this may be done, and we can as a result calculate *absolute values of entropy* for individual compounds.

5.2.1 The Third Law of Thermodynamics

The best organized physical state of a compound is that of a single perfect crystal, because each atomic or molecular unit shares a definite spatial relationship with all the others. Moreover, at low temperatures, there are fewer populated energy levels, and there is correspondingly less uncertainty about the energy of any particular unit. Translational energy is lost on crystallization, rotational energy does not persist for long, and finally, at a sufficiently low temperature, only vibrational motions remain. *At absolute zero, it is generally accepted that only the ground or lowest vibrational level would be occupied.* All units are vibrating, but they each have the smallest possible, or 'zero point' energy. The third law may be expressed:

All truly perfect crystals at absolute zero temperature have zero entropy.

There are, however, several ways in which a crystal may fail to be truly perfect. Isotopic mixtures of, say, ^{35}Cl and ^{37}Cl give rise to an entropy of mixing; different combinations of nuclear spin, as occur in ortho- and para-hydrogen, will cause some randomization. Crystals are seldom perfect, and there are other effects which must be accounted for. This will be mentioned again in Section 5.4.2.

Let us now calculate an entropy in absolute terms. Equation 5.5 may be written

$$S_{T_2} - S_{T_1} = \int_{T_1}^{T_2} \frac{C_P}{T}\,\mathrm{d}T. \qquad (5.5)$$

If $T_1 = 0$ K,

$$S_{T_2} - S_0 = S_{T_2} = \int_0^{T_2} \frac{C_P}{T}\,\mathrm{d}T.$$

In general,

$$\boxed{S_T = \int_0^{T} \frac{C_P}{T}\,\mathrm{d}T}. \qquad (5.7)$$

Values of C_P may be measured down to a few degrees Kelvin, but a means of extrapolation must be found for the last few degrees. The Debye theory of the heat capacity of solids gives, at low temperatures,

$$C_P = \alpha T^3 + \beta T,$$

or

$$\frac{C_P}{T} = \alpha T^2 + \beta,$$

and so values of C_P/T may be extrapolated to zero by plotting against T^2. It is then possible to carry out a graphical integration of eq. 5.7 from 0 K, and determine the absolute entropy.

Example 5.2 E. F. Westrum Jr. has published low-temperature heat capacities for uranium disulphide, US_2. Use these values to determine the absolute molar entropy at 298 K.

The data are shown in Table 5.1 together with the calculated values of C_P/T, and the graph of C_P/T versus temperature is shown in Figure 5.2.

On measuring the shaded area underneath the curve, using square-counting or paper weighing techniques, the absolute entropy of US_2 at 298 K is found to

TABLE 5.1 *Heat capacity data for uranium disulphide*

Temperature, K	C_P J/mol K	C_P/T
8.2	0.29	0.036
9.6	0.67	0.071
11.8	1.55	0.130
14.3	3.26	0.226
17.5	5.31	0.305
21.2	8.08	0.381
26.4	11.6	0.439
33	15.5	0.469
41	20.1	0.490
51	25.1	0.494
69	33.9	0.490
94	45.2	0.481
116	53.1	0.460
142	59.8	0.423
171	64.9	0.381
207	69.0	0.335
244	71.8	0.293
279	73.6	0.264
312	74.9	0.238

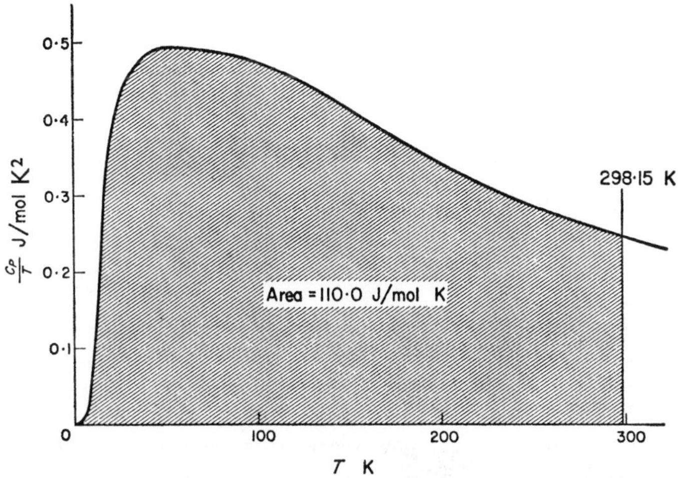

FIG. 5.2 Heat capacity and temperature graph for determination of absolute entropy of uranium disulphide.

be **110.2 J/mol K.** This result can well be verified by the reader using the data provided.

5.2.2 ΔS FOR PHASE CHANGES

Uranium disulphide is a particularly simple case, because it is solid throughout the experiment, and exists in only one crystalline form. Many compounds, however, undergo solid modifications or phase changes. These changes are easily catered for, because they occur at fixed temperatures. Thus for fusion

$$\Delta S_{\mathrm{f}} = \int \frac{\mathrm{d}q_{\mathrm{rev}}}{T_{\mathrm{f}}} = \frac{\Delta H_{\mathrm{f}}}{T_{\mathrm{f}}}. \qquad (T, P \text{ constant})$$

In general, for any phase change

$$\Delta S = \Delta H/T. \qquad (T, P \text{ constant}) \quad (5.8)$$

Using the third law, absolute entropies of any substance in its standard state, at 298 K, written S^0_{298}, may be determined from thermal measurements, using equations 5.5 (heat capacities) and 5.8 (enthalpies and temperatures of transition) as appropriate.

Example 5.3 Data are available for isobutane, from 12.53 K upwards. It does not undergo any solid/solid transitions, but melts at 113.74 K ($\Delta H = 4541$ J/mol) and boils at 261.44 K ($\Delta H = 21.29$ kJ/mol). The different stages in the calculation of S^0_{298} are summarized below.

	ΔS, J/mol K
(a) $\int \frac{C_P}{T}\,\mathrm{d}T$ for solid, 0 K - T_{f}, including Debye extrapolation below 12.53 K	68.46
(b) Fusion, $\Delta H/T = 4541/113.74$	39.93
(c) $\int \frac{C_P}{T}\,\mathrm{d}T$ for liquid, T_{f} - T_{b}	92.68
(d) Boiling, $\Delta H/T = 21\,290/261.44$	81.45
(e) $\int \frac{C_P}{T}\,\mathrm{d}T$ for gas, T_{b} - 298.15 K	12.17
$S^0_{298} =$	**294.7**

It is helpful to represent the entropy changes of isobutane pictorially, as in Fig. 5.3. Processes (a) and (c) increase the temperature, and the entropy increases because there is available a wider range of possible molecular energy states. Processes (b) and (d) represent increases in geometrical randomness.

Figure 5.4 shows in detail how the entropy of hydrogen, and of boron, increases with temperature. Notice particularly how great is the contribution due to vaporization. Crystalline boron is very highly ordered, and has a correspondingly low entropy.

Values of S^0_{298} are obtained from heat capacity and other measurements, and Table 2.1 includes a selection of such data for a number of compounds. From

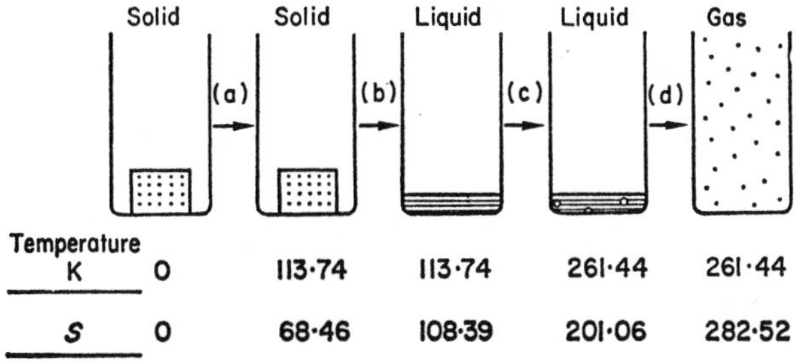

FIG. 5.3 The entropy of one mole of isobutane, C_4H_{10}, is represented pictorially. The crystal at 0 K is assumed to be perfect in every respect. Processes (a) and (c) represent temperature increases, processes (b) and (d) phase changes, and the entropy rises during each.

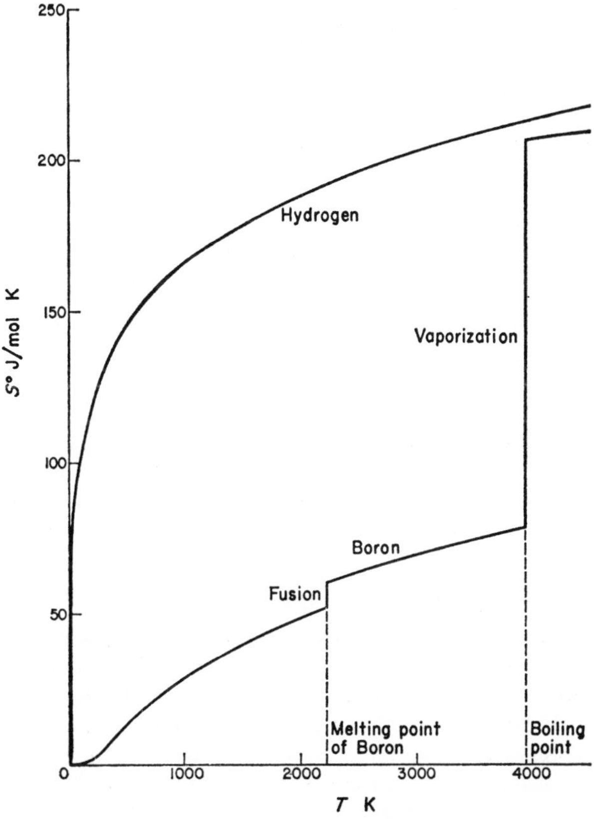

FIG. 5.4 The standard molar entropies of boron and hydrogen from 0 to 4500 K. Note the large increase on vaporization. (Data from *JANAF Thermochemical Tables*, Dow Chemical Company, Midland, Michigan).

such tabulations, it is possible to determine values of ΔS^0_{298} for any specified reaction. As usual, one can use an equation similar to eq. 1.1,

$$\Delta S^0_{298} = \Sigma S^0_{298}\,(\text{Products}) - \Sigma S^0_{298}\,(\text{Reactants}). \qquad (5.9)$$

Example 5.4 What is ΔS^0_{298} for the formation of isobutane from its elements? The reaction is

$$4C\ (\text{graphite}) + 5H_2\ (g) \longrightarrow C_4H_{10}\ (g).$$

Using values from Example 5.3 and Table 2.1, we find

$$\Delta S^0_{298} = 294.7 - 4(5.7) - 5(130.6)$$
$$= \mathbf{-381.1\ J/mol\ K.}$$

5.3 The Direction of Time

It has been said that increasing entropy is the arrow of time; but what is the connection between them? All irreversible, that is natural, processes lead to an increase of entropy *with respect to time.* If we build a large 'house of cards', this represents a very unlikely assemblage; there is only one house arrangement but a multitude of disorganized arrays. All too easily a small vibration causes reversion to a disordered state. This occurrence may cause minor dismay, but on the other hand to watch the cards spontaneously jumping and manoeuvring themselves into a perfect structure would suggest either mental derangement *or that time was going backwards.*

Left on its own, a bouncing ball will eventually come to rest. A small ball-bearing dropped onto a sheet of plate glass will rebound almost to its previous height each time, and the various changes are almost reversible. Eventually, however, the kinetic (or gravitational or distortional) energy will be converted to thermal energy, which is simply a *randomized kinetic energy operating on a molecular scale.* The motionless ball will be a little hotter than before; the molecules will be jiggling all in different directions with varying energies. That all these molecules should spontaneously organize themselves into co-ordinated upward movement, so that the ball 'jumps' up once again, is not impossible in terms of energy conservation. *It is, however, extremely unlikely*, because there is only one type of net upward movement compared with the myriads of **equally likely** schemes of random movement. The vast number of possible schemes of random movement is related closely to the greater entropy of the disordered condition.

5.4 A Statistical Approach to Entropy

It is possible to relate, mathematically, the entropy of a material to its degree of disorder. This involves a detailed analysis of the distribution of all the atoms and molecules in the various energy levels available to the system. For any particular set of conditions (temperature, volume, and so on) or *macrostate*, there is a large *number of compatible microstates*, as discussed in the previous section. This number is denoted by ω, the so-called *multiplicity* of the macrostate.

5.4.1 The Boltzmann Entropy Equation

By discussing the Joule expansion experiment in statistical terms, we shall come to accept a relationship linking ω with entropy.

First refer to Fig. 4.2; the Joule experiment may be summarized for a mole of gas which doubles its volume.

$$\text{Gas } (T, V, 1 \text{ mole}) \longrightarrow \text{Gas } (T, 2V, 1 \text{ mole}).$$

Let B denote the multiplicity of state 1. State 2 involves a greater multiplicity, because each molecule might find itself in the left or the right container. All these individual choices are compounded for the gas as a whole, so for an Avogadro's number of molecules we may write

$$\omega_1 = B; \quad \omega_2 = 2^N B$$

where $N = (6.02)10^{23}$. Thus by increasing the volume by a factor of two, the multiplicity is increased hugely. Boltzmann suggested a logarithmic relationship for entropy and multiplicity;

$$S = \frac{R}{N} \ln \omega \tag{5.10}$$

where R is the gas constant. (R/N is often written as k, the Boltzmann constant.) From this we find for the two states, that

$$\Delta S = S_2 - S_1 = \frac{R}{N} \ln 2^N = R \ln 2.$$

This expression is found to be identical with eq. 5.3 where the volume ratio V_2/V_1 is two. The Boltzmann equation finds general support in this and many other instances.

The support lent by the third law is very impressive. In a perfect crystal at 0 K, each molecular or atomic unit is indistinguishable and has an identical energy and environment. There is one and only one scheme for such a crystal, and so $\omega = 1$, and

$$S = \frac{R}{N} \ln 1 = 0.$$

In such a case, therefore, the entropy is in fact found to be zero. The third law is thus given full support by eq. 5.10.

5.4.2 Spectroscopic Entropies

Using methods of *statistical mechanics*, it is possible to calculate absolute entropies from information on electronic, vibrational and rotational energies derived from spectral measurements. It is interesting to compare such values with 'third law' entropies, which are based on the assumption that perfect order exists at 0 K. This may not always be the case, as we shall now see.

Table 5.2 shows values of entropies from both sources.

Agreement for the first four species is good. On the other hand, both hydrogen and carbon monoxide would appear to have a finite entropy at absolute zero. Can this be?

In the case of hydrogen, it appears that some disorder due to the existence of nuclear spin and of ortho- and para- forms is present at 0 K.

The carbon monoxide anomaly may be more simply explained. The molecule has almost perfect end-to-end symmetry, and so even at the lowest available

TABLE 5.2 *Comparison of third-law molar entropies with spectroscopic values* (J/mol K)

Species	$(S_{298} - S_0)$ from C_P	Spectral entropy at 298 K
Ne	146.4	146.2
A	154.8	154.8
CH_4	185.4	185.6
HCl	186.2	186.8
H_2	124.4	130.7
CO	193.3	197.9

temperature there is a random sequence of OC and CO orientations. Each molecule can thus be in one of two positions, and so, for one mole or N molecules,

$$S = \frac{R}{N} \ln 2^N = R \ln 2 = \mathbf{5.9\ J/\ K.}$$

This is the entropy of the otherwise perfect crystal of CO at 0 K, and it accounts very well for the discrepancy of (197.9 — 193.3) = 4.6 J/ K shown in Table 5.2.

PROBLEMS

5.1 Decide by inspection whether the following reactions will have values of ΔS which are positive, negative, or about zero. Check values from tables.

$$2\ H_2\ (g) + C_2H_2\ (g) \longrightarrow C_2H_6\ (g)$$
$$LiH\ (c) + H_2O\ (l) \longrightarrow LiOH\ (c) + H_2\ (g)$$
$$C\ (graphite) + CO_2\ (g) \longrightarrow 2\ CO\ (g)$$
$$NO\ (g) + O_3\ (g) \longrightarrow NO_2\ (g) + O_2\ (g)$$
$$C_2H_4\ (g) + Br_2\ (l) \longrightarrow CH_2Br.CH_2Br\ (l)$$

5.2 A 2-kilowatt electric heater runs for an hour and a half in a room at 20°C. The filament temperature is 800°C. What is the overall increase in entropy in joule/ K?

5.3 Two moles of pure supercooled water at 265 K freeze to give ice at the same temperature. At 273 K, $\Delta H_{fusion} = +6.008$ kJ/mol, and C_P (ice) = 36.8 J/mol K. Calculate ΔH and devise a reversible route to calculate ΔS, for this process.

5.4 Values of C_P for silane increase steadily with temperature:

T K	200	250	300	350	400
C_P J/mol K	35.5	39.2	43.0	47.0	51.5.

Determine the increase of entropy $(S_{400} - S_{250})$ on heating one mole of silane from 250 — 400 K, at 1 atmosphere; use a graphical method.

5.5 W. T. Smith and co-workers have published (*J. Am. chem. Soc.* **75,** 5785 (1953)) low-temperature heat capacity data for metallic rhenium.

T K	20.39	24.89	31.74	34.66	39.98	48.89	62.26	70.82
C_P, J/mol K	0.661	1.318	2.61	3.35	4.540	7.544	11.42	13.12
T K	80.05	92.88	103.04	121.82	148.41	202.65	244.81	300.01
C_P, J/mol K	14.91	17.23	18.39	20.43	22.27	24.28	25.45	25.73

Determine S^0_{298}(Re) graphically, using a $C_P = AT^3$ extrapolation at very low temperatures, where A is a constant.

5.6 The hexafluorides of tungsten, osmium, and iridium all undergo solid-solid transitions at or a little below 0°C. The process β (low temp. form) $\longrightarrow \alpha$ (high temp. form) is endothermic in each case:

	WF_6	OsF_6	IrF_6
$\Delta H_{\text{transition}}$, kJ/mol	5.86	8.24	7.11
T K	265.0	272.8	273.6

Calculate the increase of entropy for these transitions.

5.7 The spectroscopic entropy of N_2O (220.0 J/mol K) is slightly greater than the value determined by heat capacity measurements (215.2 J/mol K). A similar situation exists for certain 1-alkenes, for example

$$\text{1-duodecene, } CH_3(CH_2)_9CHCH_2,$$

and for a similar reason. Why might this be?

6

Free Energy, The Arbiter

In this chapter, we come to the most important single concept in chemical thermodynamics, that of Gibbs Free Energy, *G*. It is a function of state which provides the criterion for deciding whether or not a change of any kind will tend to occur. Physical and chemical changes will have simple numerical values of ΔG such that a negative number indicates spontaneity, a positive number tells us that the reverse change tends to occur, and a value of zero implies an equilibrium situation.

6.1 Processes in Isolated Systems

We recall that whenever a process occurs in a natural or everyday situation it does so irreversibly, and the overall change of entropy, that is of the system and surroundings together, is positive. This, then, is a general criterion of the spontaneity or otherwise of a change. It is very difficult, however, to carry out experiments in which changes occurring both in the reaction vessel (the system), and in the surroundings, are followed, because the surroundings are so ill-defined. Progress is often made when a problem is reduced to its simplest form, and here it is helpful to isolate the experiment in order to leave the surroundings unchanged. First suppose we keep the total volume constant, and use a thermally insulated vessel. (Afterwards we can extend our activities to include constant pressure processes.)

Consider the terms ΔU, q_{to}, and w_{on} under these conditions of isolation. Insulation implies that $q = 0$, and constant volume requires that the work done by expansion is also zero. Therefore $\Delta U = q + w$ must itself be zero.

First consider a small (60 ml) container of 0.03 mole of gas under high pressure, suspended in a 10-litre evacuated vessel, which is vacuum insulated. The little bottle breaks on falling, and the space above it is immediately filled with gas. There is no temperature change, but the gas is obviously much more spread out. To determine by how much the entropy has increased, we must imagine the same overall change to be carried out reversibly, as described in Section 5.1.2(a).

By applying eq. 5.3, $\Delta S = nR \ln V_2/V_1$, we have in this case,

$$\Delta S = 0.03(8.314) \ln \frac{10\,000}{60} = 1.26 \text{ J/ K}.$$

As a second example, we shall consider another experiment in which the process is quite spontaneous, and irreversible. It results, therefore, in an entropy increase, but here the visual concept of randomization is difficult to apply. A saturated hot solution of sodium thiosulphate is cooled slowly, and if care has been taken to keep the materials clean and dust-free, supercooling by many degrees is possible. The solution is then sealed in a Dewar flask. On dropping a tiny seeding crystal through a hole in the lid (Fig. 6.1) crystallization occurs,

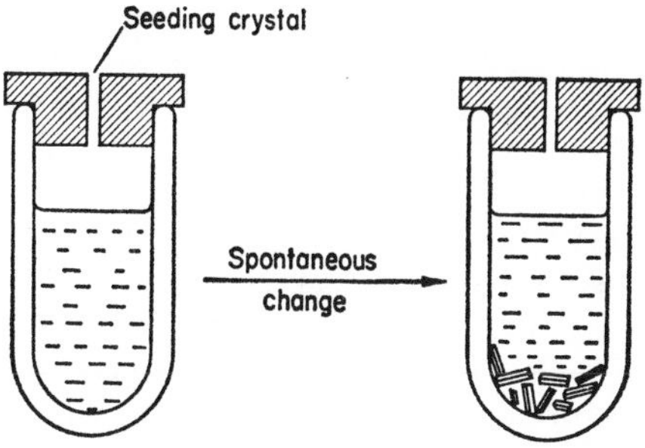

FIG. 6.1 Spontaneous organization? Precipitation of sodium thiosulphate from a super-saturated solution. Entropy in fact increases because of a temperature rise.

with an apparent increase of organization! This story tells us very forcefully that visual disorganization and entropy are not synonymous, for although highly regular crystals have been formed, the spontaneity of the process requires increased entropy. This means that the number of equivalent arrangements of microstates must be higher than before. Nevertheless, those associated with position appears to have decreased in number due to the clustering of thiosulphate and sodium ions into crystals. The volume change is negligible, and cannot account for this apparent anomaly. What other possibilites are there? The missing factor is of course the temperature, which increases. (ΔU for the *system* can still be zero, because new materials are formed, and each has its own contribution to make to the total internal energy.) This makes available a greater spectrum of thermal microstates, and so the entropy increases after all.

To summarize: $\Delta U = q + w = 0$ (System isolated),

$\Delta S_{\text{surr}} = 0$ (System isolated),

and ΔS_{system} is positive (Irreversible process, system isolated).

In both of the foregoing examples, we have seen that an increase of entropy is a valid criterion for the spontaneity of a process. In addition the second example shows that it is often misleading to interpret entropy simply as 'mixed-upness'. When we move over to normal laboratory conditions, we will find it even more difficult to keep track of what is going on. The introduction of Gibbs free energy, which simplifies the situation, will be most welcome.

6.2 Gibbs Free Energy, *G*

In our search for a criterion of whether or not a particular change would occur, we have found that in some cases at least, the attainment of minimum energy seemed to be a useful indication. The demolition of a factory chimney always leads to the same predictable result, the attainment of minimum gravitational energy. Similarly a soap bubble will quickly adjust itself to a spherical shape so as to have a minimum surface area, and hence minimum surface energy. The exothermic nature of a reaction, where energy is given out as heat, and the materials achieve lower energy, seems to be only a rough guide at the best of times. We do know that reactions which are strongly exothermic give very stable products, such as the formation from their elements of Al_2O_3 ($\Delta H^0_{f,298} = -1676$ kJ/mol), Fe_2O_3 (−822.2), and $CaCO_3$ (calcite, −1207). Several exceptions to this rule of thumb for processes of small exothermicity were mentioned in Chapter 4. In these cases it was shown that at constant temperature the increase of entropy of the system seemed to be playing a dominant part. We shall see how this factor also can be assessed.

Under isothermal conditions, these two driving forces, energy and entropy, act sometimes in concert, sometimes at variance with one another. It is none the less possible to define a new quantity, the Gibbs free energy, which enables us to arbitrate between the sometimes conflicting factors of energy seeking a minimum, on the one hand, and entropy seeking a maximum, on the other. The **Gibbs free energy** is defined as

$$\boxed{G = H - TS}\,. \tag{6.1}$$

It should be noted that H, T, and S are all properties of state. This means that G must also be a property of state. In order to appreciate the usefulness of Gibbs free energy, we shall now fully differentiate this equation, to give

$$\mathrm{d}G = \mathrm{d}H - T\,\mathrm{d}S - S\,\mathrm{d}T.$$

The majority of processes take place at constant temperature, and under these conditions $\mathrm{d}T = 0$.

Thus $$\mathrm{d}G = \mathrm{d}H - T\,\mathrm{d}S. \qquad (T \text{ constant}) \ (6.2)$$

If the pressure is constant (usually one atmosphere) we may write $\mathrm{d}q$ for $\mathrm{d}H$, giving

$$\mathrm{d}G = \mathrm{d}q - T\,\mathrm{d}S. \qquad (T, P \text{ constant})$$

We are now in a position to appreciate how G may be used as a criterion for spontaneity. We need to bring in the second law statement, which may be summarized as

$$\left|\begin{array}{l}\mathrm{d}S = \mathrm{d}q_{rev}/T \\ \mathrm{d}S > \mathrm{d}q_{irrev}/T.\end{array}\right.$$

Thus $T\,\mathrm{d}S$, being a property of state, will always be equal to $\mathrm{d}q_{rev}$. If our process is performed reversibly, that is, in a continuous state of equilibrium, then $\mathrm{d}G = \mathrm{d}q_{rev} - \mathrm{d}q_{rev} = 0$. On the other hand, a spontaneous, and therefore irreversible process will have $\mathrm{d}q_{irrev} < T\,\mathrm{d}S$, and therefore $\mathrm{d}q_{irrev} < \mathrm{d}q_{rev}$. Then, $\mathrm{d}G = \mathrm{d}q_{irrev} - \mathrm{d}q_{rev}$, and $\mathrm{d}G$ is a negative quantity. This is what we were looking for. Gibbs free energy, G, *decreases as long as the process is spontaneous* ($\mathrm{d}G$ is negative). As soon as a state of equilibrium is reached, G comes to the bottom of the hill, and reaches a minimum value ($\mathrm{d}G = 0$), which specifies that no further reaction can take place.

For constant temperature processes, eq. 6.1 may be used to apply to the initial state (reactants) and the final state (products), so that

$$G_{products} = H_{products} - TS_{products}$$
$$G_{reactants} = H_{reactants} - TS_{reactants}$$

and so $$\boxed{\Delta G = \Delta H - T\Delta S}\,. \qquad (T \text{ constant}) \ (6.3)$$

The delta has the usual significance of 'products minus reactants'.

If we use this new quantity, the Gibbs free energy *change*, ΔG for a process, we are relieved of the difficulty of following changes in the surroundings. Values of ΔH and ΔS for the process are sufficient to tell us whether the process is spontaneous or not. If ΔG is a negative number, the process will tend to occur; if zero, equilibrium is attained; and if positive, the reverse process is spontaneous. Remember that spontaneity implies only a *tendency* to occur, and that the rate of reaction may well be undetectably slow. This is because thermodynamics is concerned only with the initial and final states, whereas questions of collision rates and activation energy are the province of chemical kinetics. Thus ΔG for the reaction

$$H_2O_2\,(l) \longrightarrow H_2O\,(l) + \tfrac{1}{2}\,O_2\,(g)$$

is devastatingly negative ($\Delta G = -116.7$ kJ/mol), but hydrogen peroxide may none the less be kept, if pure, for long periods. The presence of a catalyst, such as platinum gauze, will quickly resolve any doubts about its thermodynamic instability.

Before we discuss some specific applications of Gibbs free energy, it is necessary to interpret ΔG as the greatest *useful work* available from a process at constant pressure. This will be helpful later on, for example in the chapter on electrochemistry.

6.3 Gibbs Free Energy and Maximum Work

Useful work could have many interpretations, depending on the circumstances. Here it is taken to mean any work other than the work of gaseous expansion, which is an inconvenient form. (Expansion against atmospheric pressure, $P\Delta V$, often occurs however, and must be accounted for). Electrical work is probably the most convenient (w_{elec}). From the First Law,

$$\Delta U = q_{\text{to}} + w_{\text{on}}.$$

Ignoring all forms of work other than those already mentioned,

$$\Delta U = q - P\Delta V + w_{\text{elec}}.$$

Thus $$w_{\text{elec}} = \Delta U + P\Delta V - q$$

$$= \Delta H - q. \qquad (P \text{ constant})$$

We should note here that although ΔH depends only on the difference between initial and final states of the system, q will vary with the reversibility or otherwise of the process.

First, if reversibly done, the Second Law tells us that $dq_{\text{rev}} = T\,dS$, and that if temperature is constant this may be scaled up to give

$$q_{\text{rev}} = T\Delta S.$$

Thus $$w_{\text{elec, rev}} = \Delta H - T\Delta S.$$

But from eq. 6.3, $$w_{\text{elec, rev}} = \Delta G.$$

Thus the more negative ΔG is the greater will be the useful work (electrical work, in this case) obtainable from the process.

If the process is *irreversible*, $q_{\text{irrev}} < T\Delta S$, and therefore $\Delta H - q$, equal to w_{elec}, will have a smaller value. Thus, for a spontaneous change, w_{elec} will be less than the best available, which is ΔG. We will see later that the maximum electrical work available from an electrical cell will be obtained under reversible conditions, where the cell e.m.f. is opposed by an infinitesimally smaller potential. The electrons are made to work their passage around the external circuit to the maximum of their ability. Under these conditions, the electrical work depends on the equilibrium voltage, E, and on the number of electrons made to go through the circuit, corresponding to nF coulombs; F is a unit of charge, the Faraday = 96 487 coulombs/mol of electrons and n = number of moles of electrons or 'equivalents'. This is expressed as

$$w_{\text{elec, rev}} = -nFE \text{ joules}$$

(the mole is used here in its most general form, as an Avogadro's Number of particles).

6.4 Some Processes in Terms of Gibbs Free Energy

We have now come to the stage where we can analyse phenomena occurring daily around us, in terms of the two directing influences which operate under isothermal conditions. We shall see that the two are not always acting in the same sense, and in such cases it is necessary to use the equation

$$\Delta G = \Delta H - T\Delta S \qquad (T \text{ constant}) \quad (6.3)$$

in order to decide whether or not occurrence is possible. Before going further, some general comments are in order.

First we must find numerical values for ΔH and ΔS. The enthalpy change will be derivable from the enthalpies of formation of the participants, from Table 2.1. Values of S^0_{298}, from the same table, may be similarly combined, as outlined in Section 5.2.2, to give ΔS^0.

Secondly, we should note from eq. 6.3 that whatever the sign and magnitude of ΔS may be, it will have increasing importance at high temperatures. Thus for an isothermal process carried out at low temperature, the ΔH term is dominant. At absolute zero, both ΔS and T are zero, and so $\Delta G = \Delta H$. On the other hand, many stellar processes are dominated by entropy considerations alone. We on earth, however, are neither in the freezer nor in the fire, but somewhere between the two. As a result, we must learn to cope with both factors.

6.4.1 Adsorption Processes

Clean surfaces often take up gaseous material, which is *adsorbed* onto the surface of the solid. For example, many metals adsorb oxygen, hydrogen, and water vapour, and charcoal is well known for its adsorbent properties. Heterogeneous catalysis involves adsorption of the reacting materials, in which case modification of the adsorbed species occurs, which makes further reaction that much more likely, so speeding up the overall reaction rate.

The enthalpy change for such processes is $\Delta H = \Delta G + T\Delta S$. The term ΔG is necessarily negative, if the adsorption does in fact occur; but what can be said of ΔS? In all cases of *physical adsorption*, where only weak molecular forces are involved, the process is similar to simple condensation; the molecules have less freedom on the surface, and ΔS *is negative*. In the case of *chemical adsorption*, chemical bonds are involved, and an adsorbed molecule can become dissociated. In all but a very small number of cases, such dissociation is insufficient to render the overall entropy change positive. (The adsorption of hydrogen molecules onto a sulphur-contaminated iron surface is one of the few examples where ΔS is positive.) We can now state that all adsorptions with negative entropy change, which comprise all physical and the great majority of chemical adsorptions, are exothermic. This conclusion follows directly from eq. 6.3.

6.4.2 Evaporation Phenomena

When a liquid boils, the vapour pressure is equal to the atmospheric pressure, and at any moment when the liquid is at the boiling temperature, vapour at one atmosphere pressure is in equilibrium with the liquid. Boiling liquid nitrogen may be considered as an example.

$$N_2 \text{ (liquid, 1 atm)} \underset{}{\overset{77{\cdot}3\text{ K}}{\rightleftharpoons}} N_2 \text{ (gas, 1 atm)}.$$

Provided there is no superheating, the process is reversible, and $\Delta G = 0$. This means that $\Delta H - T\Delta S = 0$, and so we may calculate the entropy change involved. At the boiling point T_B,

$$\Delta S = \frac{\Delta H_{\text{vaporization}}}{T_B} = \frac{5506 \text{ J/mol}}{77.3 \text{ K}}$$

$$= \mathbf{73.6\ J/mol\ K.}$$

If any further heat is added to the system because the Dewar flask is imperfect, the liquid temperature will rise very slightly above the boiling point. As soon as this occurs, $\Delta G = \Delta H - T\Delta S$ becomes slightly negative, and the system redresses the imbalance by evaporating further. This then cools the liquid to the equilibrium temperature.

Entropies of vaporization at the boiling points and at atmospheric pressure turn out to have values near 88 J/mol K. This was investigated empirically by Trouton, and the 'rule' named after him followed.

Table 6.1 *Some Entropies of Vaporization*

Substance	Boiling point, K	ΔH_{vap}, kJ/mol	ΔS_{vap}, J/mol K
N_2	77.3	5.69	73.6
$(C_2H_5)O$	308	27.2	88.3
C_6H_6	353	31.4	88.7
$C_6H_5.CH_3$	384	33.4	87.0
Hg	630	59.4	94.6
H_2O	373	40.6	108.8

In most cases, the entropy increase is due only to the isothermal randomization of the molecules from the liquid to the vapour at one atmosphere. In the case of water, because there is significant hydrogen-bonding in the liquid the entropy of liquid water is smaller than it would otherwise be. As a result, the relative increase in elbow-room which the water molecules 'feel' on evaporation is greater, and this is reflected in the high value of 108.8 J/mol K for the entropy change.

Many millions of gallons of water are lost annually from lakes and rivers the world over from spontaneous evaporation. In regions in Australia where water conservation and control are important, monomolecular layers of fatty acids and similar film-forming materials are used to limit evaporation. Why, and to what extent is this evaporation, which takes place well below the boiling point, spontaneous? First, let us assume that a dry air is blowing over the lake. At a temperature of 30 °C the water has a vapour pressure of 31 mmHg. The latent heat of evaporation is roughly +40.6 kJ/mol, which corresponds to a highly endothermic process. At the boiling point of water, the entropy term $T\Delta S$ = 373(108.8)/1000 = 40.6 would be just sufficient to make ΔG marginally negative; but the temperature of the Australian outback is a mere 30 °C (303 K).

The situation could be expressed like this:

$$\begin{array}{ccccc} \Delta G & = & \Delta H & - & T\Delta S. \\ \text{(probably small, but negative)} & & +40.6 & & (303)(?) \end{array}$$

This provides a value for ΔS near 40 600/303 = 134 J/mol K. This means that for the process to occur the entropy change for the process

$$H_2O\ (\text{liq., 1 atm, } 30\,°C) \longrightarrow H_2O\ (\text{dilute vapour in air, 1 atm, } 30\,°C)$$

must be larger than 134 J/mol K. This value, significantly larger than the value for ΔS_{vap} quoted above, can easily be achieved, but only by producing a *sufficiently dilute* vapour.

Because spontaneous evaporation into a dry atmosphere causes cooling of the remaining water, the damp clothes on the washing line freeze in cool weather, and early peoples were afforded the equivalent of modern refrigeration. The early Egyptians, living in the exceptionally dry desert air, were probably quite familiar with ice manufactured by this method.

6.4.3 Endothermic Chemical Processes

The previous examples have been mainly physical processes, but chemical ones are equally amenable to analysis in terms of free energy. The reaction which comprises dehydration by thionyl chloride was mentioned previously. It is endothermic, but proceeds with great certainty none the less. It would seem probable that ΔG is fairly large in addition to having a negative sign. The reaction may be represented as

$$SOCl_2\ (l) + H_2O\ (l) \longrightarrow SO_2\ (g) + 2HCl\ (g).$$

The perhaps surprising spontaneity of this reaction is closely linked to the fact that copious gaseous products are given off by purely liquid reactants. The entropy will be seen to increase markedly. For this process, ΔS^0_{298} is easily determined from values of S^0_{298} obtained from Table 2.1.

$$\begin{aligned} \Delta S^0_{298} &= \sum_{\text{PRODUCTS}} S^0_{298} - \sum_{\text{REACTANTS}} S^0_{298} \\ &= S^0(SO_2) + 2S^0(HCl) - S^0(SOCl_2) - S^0(H_2O) \\ &= 248.1 + 2(186.8) - 215.7 - 69.9 = \mathbf{+336.1\ J/mol\ K.} \end{aligned}$$

The value for ΔH^0_{298} for the same reaction is calculated from the standard enthalpies of formation of the various chemicals, also to be found in Table 2.1.

$$\begin{aligned} \Delta H^0 &= \Delta H^0_f(SO_2) + 2\,\Delta H^0_f(HCl) - \Delta H^0_f(SOCl_2) - \Delta H^0_f(H_2O) \\ &= \mathbf{+49.4\ kJ/mol.} \end{aligned}$$

Then, at 298 K,

$$\Delta G^0 = \Delta H^0 - T\,\Delta S^0$$
$$= +49.4 - 298(+336.1)/1000$$
$$= \mathbf{-50.8\ kJ/mol\ SOCl_2.}$$

It is important to realize the full meaning of ΔG^0 consequent upon our using standard state data for entropies (S^0) and enthalpy changes (ΔH^0). In both cases, the superscript ^{0} infers that all materials, whether reactants or products are present in their standard states, that is, in effect the stable states in which one handles them in laboratories at atmospheric pressure. The full significance of the result may be summarized by saying that for the reaction in which one mole of liquid water is in contact, and reacts with liquid thionyl chloride to produce sulphur dioxide and hydrogen chloride, *each present as a gas at one atmosphere partial pressure,* the Gibbs free energy change at 298 K is -50.8 kilojoules.

The link between Gibbs free energy and equilibrium constant can be appreciated qualitatively at this stage. The equilibrium constant in terms of pressures is given by:

$$K_P = p_{SO_2}\, p^2_{HCl}.$$

The partial pressures of the liquid reactants are invariant, and do not appear explicitly. It is apparent that a high applied pressure will discourage the reaction from proceeding further, and that a low pressure will enhance it. At a sufficiently high pressure, the system could be brought to equilibrium, and the process made reversible. Consequently, K_P is greater than one. It is intuitively feasible that a negative value of ΔG is associated with a value of K greater than unity. Conversely, positive values of ΔG would imply small values of K, usually less than one. The relationship.

$$\Delta G = -a \log K$$

would fulfil these simple requirements, where a is a constant term. In the next chapter, such a relationship will be derived exactly.

As a second example of an endothermic, but spontaneous process, we shall discuss the decomposition of cupric oxide to cuprous oxide and oxygen. The equation for the reaction and relevant data are

$$2\ CuO\ (s) \rightleftharpoons Cu_2O\ (s) + \tfrac{1}{2}\ O_2\ (g)$$
$$\Delta H^0_{298} = +145.6\ kJ/mol$$
$$S^0_{298}(CuO) = 43.5\ J/mol\ K$$
$$S^0_{298}(Cu_2O) = 100.8\ J/mol\ K$$
$$S^0_{298}(O_2) = 205.0\ J/mol\ K.$$

From this we find that $\Delta S^0 = +116.3$ J/mol K. This is therefore a reaction torn by internal conflict; the entropy change is favourable, but the energy change is strongly endothermic, and therefore unfavourable.

The actual stability of cupric oxide depends on the temperature. At 300 K, $T\Delta S^0$ is equal to $300(+116.3)/1000 = 34.89$ kJ/mol. Thus at room temperature,

$$\Delta G^0 = +145.6 - 34.89 = +\,110.7\ kJ/mol.$$

Decomposition does not tend to occur, and cupric oxide is stable. However, at approximately 1250 K, the temperature is high enough to cause ΔG^0 to change sign, and at 1300 K, $T\Delta S^0$ has the value of 151.2 kJ/mol, with the result that ΔG^0 is -5.6 kJ/mol, and decomposition is favoured.

To summarize, we may say that the sign of the Gibbs free energy change, which determines the direction of a process, depends on the two terms, ΔH^0 and $T\Delta S^0$ appearing in eq. 6.3. In many cases they reinforce one another, but occasionally they tend to nullify each other. A change of temperature might then alter the sign of ΔG^0 and therefore the feasibility of the process.

6.5 Standard Free Energy Changes

Values of ΔG^0 are required for a multitude of reactions at many different temperatures, and in general one can calculate these from the appropriate values of ΔH^0 and S^0. It would be unnecessarily tedious to tabulate data for all possible reactions; ΔG^0 values for formation reactions at 298 K are none the less very useful. These may be added or subtracted (ΔG is a function of state) to provide data for the reaction of interest, and then corrections made if the reaction temperature is higher (or lower) than the standard temperature of 298 K. (Such methods are elaborated in Chapter 10.) Table 2.1 includes a basic selection of such data; values for elements in their standard states are, of course, zero. Values for most but not all compounds are negative. The fact that values are positive in some cases does not imply that the compound (diborane, B_2H_6, for example) cannot be made. It does show, however, that the compound is 'thermodynamically unstable', and will decompose if time or catalysts are available.

PROBLEMS

6.1 Two moles of ideal gas expand irreversibly from 4 litres to 20 litres at 54°C. Calculate ΔS for this change, and hence ΔG.

6.2 Estimate the normal boiling points of the following materials, the vaporization enthalpies for which are known:

	$SnCl_4$	cyclo-C_6H_{12}	$(CH_3)_2CO$	C_2H_6
ΔH_{vap}, kJ/mol	33.05	30.06	29.08	15.94 .

6.3 Vaporization and transport of tungsten normally limit the life of the tungsten filament lamp. The *quartz-iodine* lamp contains small quantities of iodine which react with deposited tungsten:

$$\underset{\text{(on quartz envelope)}}{W\,(c)} + I_2\,(g) \longrightarrow WI_2\,(g).$$

The WI_2 vapour then decomposes and deposits tungsten on the filament. The iodine is then liberated for further scavenging.

(a) Find ΔG^0 for the above equation at 350°C, the normal temperature of the quartz envelope.

(b) Find the minimum temperature of the filament at which WI_2 will decompose.

(Assume for both parts that ΔH and ΔS are independent of temperature.)

6.4 If glass were used in place of quartz in the quartz-iodine lamp (previous question), would there be any danger of the reaction

$$\underset{\text{(in glass)}}{Na_2O} + I_2\,(g) \longrightarrow \tfrac{1}{2}\,O_2\,(g) + 2\,NaI\,(c)$$

occurring? Assume ΔH^0 and ΔS^0 are temperature-invariant. (In fact quartz is widely used. What other factors lead to this choice?)

6.5 Palladium oxide, PdO is moderately stable. Over the temperature range 298–1200 K, the standard free energy of formation is given by the approximate equation

$$\Delta G_f^0(\text{PdO (c)}) = -97\,910 + 87.0\text{T}.$$

What are ΔH^0 and ΔS^0 from this equation? Units are joules.

Does PdO become more or less stable as the temperature increases? Investigate the situation at 900 K and 1150 K.

6.6 Chlorine reacts with methane at laboratory temperatures:

$$CH_4\ (g) + Cl_2\ (g) \rightleftharpoons CH_3Cl\ (g) + HCl\ (g).$$

By determining ΔG_{1000}^0 at 1000 K from the following free energies of formation, find out if this reaction will occur at this higher temperature.

Substance	CH_4 (g)	CH_3Cl (g)	HCl (g)
$\Delta G_{f,1000}^0$,kJ	−5.816	4.477	−98.53

6.7 Researchers at the Steel and Alloy Institute, Moscow, have published the following data on the reaction

$$Si\ (c) + SiCl_4\ (g) \rightleftharpoons 2\ SiCl_2\ (g):$$

$$\Delta H_{1300}^0 = +531\ \text{kJ/mol},$$

$$\Delta G_{1300}^0 = -5.837\ \text{kJ/mol}.$$

(a) What is ΔS^0 at 1300 K? (b) Is silicon stable to $SiCl_4$ at lower temperatures? (c) The value for ΔH^0 is perhaps higher than expected. Is it reasonable in the light of bond energy arguments?

6.8 For the dissociation reaction

$$PCl_5\ (g) \rightleftharpoons PCl_3\ (g) + Cl_2\ (g)$$

at 472 K, $\Delta G^0 = +5.77$ kJ/mol, but at 524 K it has changed to −4.31 kJ/mol. What qualitative change does this indicate? Does eq. 6.3 lead you to expect such a change? Make what checks you can on these data.

7

Chemical Equilibrium

This chapter is concerned with the change of ΔG during the course of a reaction. During such a change, materials are constantly reacting and reforming, but the trend will always be in the direction for which ΔG is negative. Finally a certain combination of active masses will be reached where ΔG will be zero, and equilibrium achieved. This state is characterized by the equilibrium constant, which will have a fixed value at a given temperature. Before these calculations may be made in the simplest way, the idea of activity, which embraces the concepts of gas pressure, solution concentration, and active masses of pure solids and liquids will be cast in its unifying role. Also, the variation of free energy with change of activity, or 'active mass', will be calculated.

7.1 Preamble

We have seen that a reaction will proceed just as long as ΔG is negative, although the reaction may take place extremely slowly. Further, a sufficient criterion for the state of equilibrium, where no net reaction in either direction occurs, is that $\Delta G = 0$.

It is well known that the reaction involving the hydrolysis of ethyl acetate to ethyl alcohol and acetic acid is a so-called equilibrium reaction, in the sense that measurable quantities of all four components may be detected in the reaction mixture after an indefinite time. What is not so widely known, perhaps, is that *all* reactions are 'equilibrium' reactions, and that no reaction goes literally to completion. Even the formation of liquid water from gaseous hydrogen and oxygen is not complete, and there is a certain finite pressure of hydrogen in equilibrium with the liquid. The fact that, on average, in the air over the Atlantic Ocean only five hydrogen molecules are in equilibrium with it does not detract from the main point, that *no reaction* is complete.

Consider in more detail the hydrogenation of ethylene; the equation is

$$\underset{1\ \text{atm}}{C_2H_4\,(g)} + \underset{1\ \text{atm}}{H_2\,(g)} \overset{\text{catalyst}}{\rightleftharpoons} C_2H_6\,(g).$$

Suppose the initial partial pressures to be one atmosphere for each reactant. (The **partial pressure** of one component of a mixture of gases is defined as the pressure it would exert if it were alone in the available space. For an ideal mixture of perfect gases, the total pressure is the sum of the various partial pressures.) At 25 °C, ΔG^0 for this reaction is -101.0 kJ/mol, and so the driving force is considerable, and reaction takes place over the catalyst with great vigour. The temperature of the reaction vessel is kept constant. As the reaction occurs, the partial pressures of ethylene and hydrogen drop, and their reduced 'active mass', or 'presence', causes the value of ΔG to become a smaller negative number. This situation is aggravated by the increasing active mass of the product, ethane, and the associated back reaction. Eventually, the active mass of the reactants is so low and that of ethane so high, that the forward rate exactly equals the rate of the backward reaction and no net reaction occurs, so $\Delta G = 0$. It seems plausible that a large negative value for ΔG^0 should lead to near complete reaction from left to right, and also to a large equilibrium constant. This relationship we shall derive.

Before we can measure the changes of ΔG during the course of a reaction, we must find the link between G and pressure, concentration, and physical state.

7.2 Variation of G with Gas Pressure

We start with the definition of G,

$$G = H - TS.$$

By substituting $H = U + PV,$

we have $G = U + PV - TS.$

Complete differentiation of this yields

$$\mathrm{d}G = \mathrm{d}U + P\,\mathrm{d}V + V\,\mathrm{d}P - T\,\mathrm{d}S - S\,\mathrm{d}T.$$

However, as $\mathrm{d}U = \mathrm{d}q - P\,\mathrm{d}V,$

and $\mathrm{d}S = \mathrm{d}q/T,$ for a reversible change, we may

substitute $\mathrm{d}U = T\,\mathrm{d}S - P\,\mathrm{d}V.$

This immediately looks much more promising, as

$$\mathrm{d}G = (\cancel{T\,\mathrm{d}S} - \cancel{P\,\mathrm{d}V}) + \cancel{P\,\mathrm{d}V} + V\,\mathrm{d}P - \cancel{T\,\mathrm{d}S} - S\,\mathrm{d}T$$

or

$$\boxed{\mathrm{d}G = V\,\mathrm{d}P - S\,\mathrm{d}T}. \tag{7.1}$$

This simple equation has great subsequent application and should be memorized. For our particular need, we are interested in the effect of pressure, and can put $\mathrm{d}T = 0$ (constant temperature), so

$$\left(\frac{\mathrm{d}G}{\mathrm{d}P}\right)_T = V.$$

Before going on to integrate this relation, it is rewarding to introduce a new term, the chemical potential, which is closely related to free energy.

7.2.1 Chemical Potential, μ

We have so far spoken mainly of *closed systems*, those in which the amount of each component is unchanged. If material is added to or taken from the system, or, more important, materials are consumed or produced by a chemical reaction, then we are unable completely to specify the system at every stage of the alteration. The thermodynamicist reacts to this situation much as nature to a vacuum; the gap must be filled. In addition to temperature and pressure, which have sufficed until now to describe the system, we must now introduce a new variable into our repertoire. The amount of each material present will now figure prominently. *Chemical potential* is here defined as

$$\mu_i = \left(\frac{\partial G}{\partial n_i}\right)_{T,\,P,\,n_j,\,\ldots} \tag{7.2}$$

That is, the chemical potential of the component i is the partial derivative of G with respect to the number of moles of i, written n_i; temperature, pressure, and the amounts of all other components, n_j, etc., are held constant. It follows from this that where before (eq. 7.1) we wrote $dG = V\,dP - S\,dT$, changes in G could now arise due to changes in n_1, n_2, the number of moles of components 1, 2, etc. and we have to write

$$dG = V\,dP - S\,dT + \left(\frac{\partial G}{\partial n_1}\right)dn_1 + \left(\frac{\partial G}{\partial n_2}\right)dn_2 + \dots \qquad (7.3)$$

Here, the additional terms account for new material. Summation at constant temperature and pressure gives simply

$$G = \sum_{\text{all } i} \mu_i n_i. \qquad (T, P \text{ constant}) \ (7.4)$$

The rather cumbrous definition of μ given in eq. 7.2 will be more acceptable when it is realized that for a pure material, the free energy varies linearly with the number of moles, and that

$$\mu = \frac{G}{n}.$$

In this case, μ is merely the *molar free energy*, or free energy per mole, written sometimes with a bar, as $\bar{G}$. (To write G for a compound does not imply measurement in absolute terms. The value is the Gibbs free energy related to the elements, which are arbitrarily assigned zero values. The term G is thus identified with ΔG_f^0 for the compound.) This simple equation is all that is needed to define μ provided that the mixture behaves ideally. Unfortunately, no mixture is ideal in

TABLE 7.1 ΔH *of mixing in* 1 : 1 *ratios, to form* 1 *mole of mixture*

Mixture	Temperature, K	ΔH_{mix}, J/mol	Reference
Benzene/toluene	353	46	a
Benzene/CCl_4	298	109	b
Acetone/CS_2	308	1460	b

[a] CHEESEMAN, C. H. and LADNER, W. R. *Proc. R. Soc.* **A229,** 387 (1955).
[b] STAVELEY, L. A. K., TUPMAN, W. I. and HART, K. R. *Trans. Faraday Soc.* **51,** 323 (1955).

the sense that each molecule sees all surrounding molecules as identical neighbours. There are interactions between them, and usually A–A interactions are unlike A–B interactions, and usually these are both unlike those felt between B and B. In the few cases of near ideal mixtures, we find great chemical similarities, such as for the benzene/toluene, and ethylene bromide/propylene bromide systems. For these mixtures, ΔH and ΔV of mixing are small but non-zero. (For truly ideal solutions, ΔH and ΔV are zero.) The magnitude of these quantities is therefore a measure of the non-ideality of the mixture.

Table 7.1 shows the wide variation which exists in solution behaviour. The acetone/carbon disulphide system is far from ideal; this non-ideality will also show up in properties such as the vapour pressure curves, and the boiling point diagram. As a result of all this, we could say that μ_i for a compound i (the subscript implies a mixture) depends not only on the nature of the compound, *but also on the environment in which it finds itself.* In this more complete sense, μ_i is known as the **partial molar free energy,** and may be pictured as the change of free energy on adding one mole of the material to a very large and invariant system.

Consider two moles of pure acetone, and suppose the Gibbs free energy is G_{acet}. μ is simply $G/n = G_{\text{acet}}/2$, and the free energy of the acetone sample is 2μ. Alternatively, consider a lake as the system, to which the two moles of acetone are added. By how much does the free energy of the system increase? Temperature and pressure may be supposed constant, and, for the water molecules at least, their environment is unchanged. Acetone, however, now finds itself in a pure water environment, and μ_{acet} will be

$$\left(\frac{\partial G}{\partial n_{\text{acet}}}\right)_{T,\,P,\,n\ \text{H}_2\text{O}}$$

and this may well be different from G/n. Using eq. 7.3, which says in this case that

$$\mathrm{d}G = \left(\frac{\partial G}{\partial n_1}\right)_{T,\,P,\,n_2} \mathrm{d}n_1,$$

we have

$$\mathrm{d}G = 2\left(\frac{\partial G}{\partial n_{\text{acet}}}\right)$$
$$= 2\,\mu_{\text{acet}}.$$

(Again, we emphasize that the subscript to μ implies a mixture.)

7.2.2 Pressure and Chemical Potential for Ideal Gases

At the end of Section 7.2, we concluded that

$$\left(\frac{\mathrm{d}G}{\mathrm{d}P}\right)_T = V, \quad \text{or} \quad \mathrm{d}G = V\,\mathrm{d}P.$$

Using molar quantities, we can write instead

$$\mathrm{d}\mu = \bar{V}\,\mathrm{d}P,$$

where $\bar{V}$ is the volume of one mole. Now assume that the equation of state holds, and $\bar{V} = RT/P$. Then

$$\mathrm{d}\mu = RT\frac{\mathrm{d}P}{P}. \qquad (T \text{ constant}) \quad (7.5)$$

On integrating this between two states, A and B, we have

$$\mu_{\text{B}} - \mu_{\text{A}} = RT(\ln P_{\text{B}} - \ln P_{\text{A}})$$

or

$$\mu_{\text{B}} = \mu_{\text{A}} + RT \ln P_{\text{B}}/P_{\text{A}}. \qquad (T \text{ constant}) \quad (7.6)$$

Thus A could represent some reference state, with a chemical potential of μ_{A}. Let us denote this reference potential and pressure by theta $^\theta$. This means that we can dispense with A and B, and write

$$\boxed{\mu = \mu^\theta + RT \ln P/P^\theta}. \qquad (T \text{ constant}) \quad (7.7)$$

The choice of this reference state or 'starting point' is, of course, quite arbitrary, but the criterion of ease of working is usually adopted. The most generally useful reference chemical potential is that at a pressure of one atmosphere. This will be denoted by a 'superscript zero', so we now have

$$\boxed{\mu = \mu^0 + RT \ln (P(\text{atm})/1 \text{ atm})} \,. \quad (T \text{ constant}) \; (7.8)$$

The physical reality of this equation is seen best in terms of a table of $(\mu - \mu^0)$ values at various pressures (Table 7.2).

TABLE 7.2 *Changes of chemical potential resulting from non-unit pressures expressed as* $(\mu - \mu^0)$, *for an ideal gas at* 298 K

Pressure, atmospheres	Change of chemical potential, $(\mu - \mu^0)$, kJ
0.01	−11.42
0.05	−7.40
0.1	−5.69
0.5	−1.72
1	0
2	+1.72
10	+5.69
20	+7.40
100	+11.42

Before moving on to real, and therefore non-ideal gases, let us take stock. Equation 7.8 shows us that the chemical potential, for a pure gas, varies with pressure; that μ^0 is a reference value, and that the logarithmic term allows for non-unit pressure. A dimensional check reveals that the logarithmic term has no units, and that μ has the units of RT, that is joules per mole. So far so good.

7.2.3 Chemical Potential for Real Gases

Of course, no gas is truly ideal; real gases follow the $PV = RT$ relationship to a greater or lesser extent, but are unable to follow it exactly because of their physical constitution. Molecules interact with each other, and they have a volume of their own. None the less, real gases come very near to ideality at low pressures, and it can be said that the equation of state is a 'limiting law', that is, for one mole,

$$\lim_{P \to 0} (PV) = RT.$$

Many industrial processes occur only at high pressures, and so we must come to grips with this problem of non-ideality.

In formulating eq. 7.5, the assumption of ideality was made (unjustly as it turns out) and $\bar{V}$ put equal to RT/P. Therefore, eqs 7.7 and 7.8, which follow from it, also have the 'assumption of perfection' within them, and they too will therefore fail to describe a *real* gas. There are now two courses of action which we could take. Firstly, we could substitute for $\bar{V}$ a more complicated expression derived from the Van der Waals equation, or another empirical equation of state. This would be a lengthy undertaking, and as new data became available fresh integrations would have to be performed. The second alternative would be to accept eq. 7.7 on one condition; that in place of P we put a new quantity *which revalidates this relationship*. Hopefully, it would not vary very much from the ideal pressure, and could be approximated as such when the pressure was low enough. The second alternative is adopted; on the whole, it is less onerous. The new function is called **fugacity,** which is derived from the Latin *fugere*, to flee, and means literally 'escaping tendency'. It is denoted by f. Using this definition, for a real gas, we can rewrite eq. 7.7 as

$$\boxed{\mu = \mu^{\theta} + RT \ln \frac{f}{f^{\theta}}}. \qquad (T \text{ constant}) \ (7.9)$$

A formal definition would be double-headed. Firstly eq. 7.9 is a true expression of the chemical potential of a real gas, and secondly fugacity and pressure become identical at zero pressure,

$$\operatorname*{limit}_{P \to 0} \left(\frac{f}{P} \to 1 \right).$$

Fugacity is not, repeat not, the actual pressure. It must not only account for the actual behaviour of real gases, but it must also overcome the erroneous assumptions of perfection which are still part of the fabric of the equation. It therefore over corrects. To give an example, nitrogen at 25 °C and 1000 atmospheres is well away from ideality. The ideal pressure ($= nRT/V$) is 498 atm, but the fugacity is in fact 1839 atmospheres.

7.2.4 Activity

The situation is streamlined somewhat by substituting an abbreviated form of the fugacity ratio f/f^{θ}. We define the **activity,** a as $a = f/f^{\theta}$. As a result, eq. 7.9 can be made to look much more attractive. That is,

$$\mu = \mu^{\theta} + RT \ln a.$$

This equation may be expressed in words. 'The chemical potential is equal to a reference value μ^{θ}, associated with a fugacity f^{θ}, plus the correction term $RT \ln f/f^{\theta}$. The ratio f/f^{θ} is abbreviated a.' In many cases but not all, f^{θ} is assigned a value of one atmosphere. (Note that the reference *fugacity* is one atmosphere, and that the *pressure* may well differ from this value.) In this case $a = f/f^{\theta} = f/1$. The denominator is stressed, as it shows that ***a* is dimensionless.** We have now defined a **standard state** for which a zero superscript is applicable, and can write a simplified equation, which is widely useful,

$$\boxed{\mu = \mu^{0} + RT \ln a}. \qquad (T \text{ constant}) \ (7.10)$$

Here the superscript zero implies, for real gases, a unit fugacity standard state. If, during an experiment, the fugacity happens to coincide with the standard state, then the activity is one, and the chemical potential is equal to μ^{0}.

So far, activity has been treated as a somewhat abstract parameter, but we can relate it to the actual pressure by means of an *activity coefficient* γ, when we state that

$$a = \gamma P.$$

Obviously, γ is unity in the case of an ideal gas, and some other value for a real gas. Now, eq. 7.10 takes the form

$$\mu = \mu^0 + RT \ln P + RT \ln \gamma.$$

Comparison of this with eq. 7.8 shows that the final term merely allows for non-ideality. It is no more than a correction term. (When gaseous mixtures are involved, it is the various partial pressures p_i, or partial fugacities, f_i, which become important. In such a case, f_i would represent the fugacity of the compound i in the particular environment in which it found itself.)

7.3 The Active Mass of Pure Liquids and Solids

When a pure solid or liquid takes part in a reaction, it does so to the extent that individual molecules are prepared to leave the mass of the material. This tendency is closely linked to the vapour pressure of the material, and is constant at a given temperature. The vapour pressure of a piece of nickel, say, although very small, is finite, and while this depends principally on temperature, there is a slight dependence on the *total* applied pressure.

Let us imagine a super-strong muslin bag, in which a lump of nickel is squeezed to a pressure of many hundreds of atmospheres. The metal will contract by a small amount, and the concentration of the metal may be said to increase. It is then no surprise that its vapour pressure, or fugacity, increases slightly. However, the change is negligibly small. An actual example is always convincing. Water has a vapour pressure of 0.0312 atm at 25°C and a total pressure of one atmosphere, but this increases only to 0.0643 atm at 1000 atmospheres total pressure. Except for very high pressures, the effect may be disregarded. If we use fugacities of solids and liquids to cope with all cases of non-ideality, we can use eq. 7.9. We choose a standard state in much the same way as before. Thus in place of f^θ is written f^0, the fugacity *at a total pressure* of one atmosphere; the corresponding chemical potential is μ^0. Then

$$\mu = \mu^0 + RT \ln f/f^0. \qquad (T \text{ constant})$$

However, we have seen that the actual fugacity will vary little with the total pressure, and so f/f^0 is very close to unity. This is another way of saying that the activity is unity, at any moderate pressure. Thus, to summarize, for pure liquids and solids, μ is virtually constant, and the materials are assigned unit activity, that is $a = 1$, and $\mu = \mu^0$.

7.4 Activity of Materials in Solution

7.4.1 Solvents

We have seen that the activity of a gas is related to its effective pressure. For liquids, it is the liquid vapour pressure which is important, and for solutions, the vapour pressure is linked to the concentration of the component of interest. The concentration of a solvent or 'major component' is conveniently measured as the mole fraction, x. For ideal vapours and an ideal solution it may be shown that the chemical potential of a component number 1 is

$$\mu_1 = \mu^0_{(\text{pure } 1)} + RT \ln x_1. \qquad (\text{ideal solution})$$

The term $\mu^0_{(\text{pure } 1)}$ represents the chemical potential of the pure material, where x_1 is unity. For non-ideal solutions, a term a, the activity of the solvent, which may be thought of as the effective mole fraction, is introduced, and we have, for component 1,

$$\mu = \mu^0{}_{(a=1)} + RT \ln a.$$

The standard state of unit activity *corresponds to pure solvent* or *unit mole fraction.* (All solutions approach ideality at zero solute concentration.) Although the standard state is different, the form of this equation is identical to that of eq. 7.10. The activity may be linked with the actual mole fraction by the activity coefficient, γ; that is $a = \gamma m$. For ideal solutions, $\gamma = 1$.

7.4.2 Solutes or Minor Components

For solutes, molality, m, defined as the number of moles of solute dissolved in 1000 g of solvent, is a more convenient measure of concentration. For any given solvent it is approximately proportional to the mole fraction. An equation similar to eq. 7.10 may be derived for solutes,

$$\mu = \mu^0{}_{(a=1)} + RT \ln a$$

where activity is a measure of the effective molality. For dilute solutions, the molality activity coefficient, γ, is unity and $a = \gamma m = m$. The standard state ($a = 1$) corresponds to the ideal state of unit molality. (Such an ideal state is hypothetical, because all solutions deviate from ideality. However, by extrapolating up from a very dilute, ideal region, the state is adequately described.)

7.5 A Summing Up; Activity as a Unifying Concept

(a) We have treated all manner of materials, solid, liquid, gas, pure and mixed.
(b) In each case, the *active mass* was identified with activity, a.
(c) A *standard state*—a starting point for each material—was assigned a value of *unit activity*.
(d) In each case, the chemical potential is given as

$$\boxed{\mu = \mu^0{}_{(\text{activity}=1)} + RT \ln a}\,.$$

(e) The standard states, with $a = 1$ in each case, are summarized as follows:

Pure gas $f = 1$ atmosphere.

Pure liquids and solids The pure materials have an activity of virtually unity over a wide range of pressures.

Gas mixtures Ideal perfect mixture, $p_i = 1$.
Mixture of real gases, $f_i = 1$.

Solid or liquid solutions For the major component, $x = 1$.
Minor component, hypothetical ideal state, $m = 1$.

7.6 Practical Aspects of Activity

Detailed tables of fugacity at different pressures are available for many gases. Luckily it is not necessary to make actual measurements for every gas (unless justified by a need for high accuracy). Most gases show similar behaviour in the region of their critical points; these similarities are most easily seen in terms of fugacity coefficients, $\gamma = f/P$, if the pressures are expressed as *reduced pressures* (P/P_c, where P_c = critical pressure), and the temperatures as *reduced temperatures* (T/T_c, T_c = critical temperature). As an example, it is found that the fugacity coefficient for methane at a reduced pressure of $P/P_c = 2$ and reduced temperature of $T/T_c = 2$ is near 0.94. This means that $f = 0.94P$. For trichlorosilane, under the corresponding conditions of $P/P_c = T/T_c = 2$, γ is also very close to 0.94. In this way, it is possible to draw up charts for all gases, showing fugacity ratio varying with reduced pressure, for some chosen value of the reduced temperature. Such a chart is shown as Fig. 7.1. Resulting corrections are small, that is f/P is near one, as long as the temperature is above twice the critical value, and the pressure below twice its critical value.

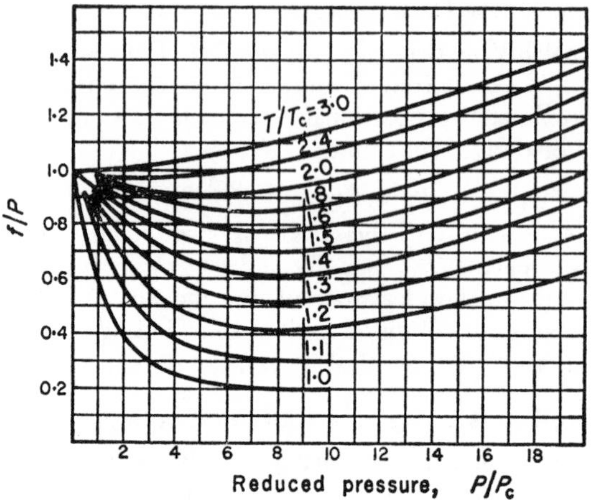

FIG. 7.1 Generalized fugacity curves for moderate pressures. The activity coefficient f/P may be found for any gas for which critical data are known. (Adapted from S. Glasstone, *Thermodynamics for Chemists*, p. 257, D. Van Nostrand (1947).)

Example of fugacity calculations Ethanol takes part in a reaction at 300°C, with a partial pressure of 30 atm. Table 10.2 gives the critical point data, which are $T_c = 516$ K, and $P_c = 63$ atm. What is the fugacity under these conditions? In this case we have reduced temperature $= T/T_c = 573/516 = 1.11$, and reduced pressure $= P/P_c = 30/63 = 0.48$. From Fig. 7.1, we find $\gamma = f/P = 0.92$. Thus $f = \gamma P = 27.6$ atm.

7.7 Equilibrium and the Reaction Isotherm

Table 2.1 includes values of ΔG^0 for a selection of formation reactions. When properly combined, these provide ΔG^0 data for a wide variety of other reactions. However, these values apply only to a *hypothetical situation* in which reactants and products are in their standard states, that is at unit activity. However, the free energy change can now be worked out for any other set of conditions, because we know how the chemical potential varies with activity. The vital relationship is of course $\mu = \mu^0 + RT \ln a$.

Consider a perfectly general reaction, such as

$$a\text{A} + b\text{B} + \ldots \rightleftharpoons p\text{P} + q\text{Q} + \ldots,$$

in which aA represents 'a moles of A', and so on. In a particular set of circumstances, we may be required to know whether or not the reaction will tend to occur. We may calculate ΔG^0 from ΔG_f^0 data, but this refers only to standard states.

Let us start with what is known. Because ΔG^0 is G^0 (products) minus G^0 (reactants), we may use eq. 7.4 to write

$$\Delta G^0 = \sum_{\text{products}} n_i \mu_i^0 - \sum_{\text{reactants}} n_i \mu_i^0$$
$$= p\mu_P^0 + q\mu_Q^0 - (a\mu_A^0 + b\mu_B^0)\ldots \quad (7.11)$$

Similarly, for non-standard states, we may write

$$\Delta G = p\mu_P + q\mu_Q - (a\mu_A + b\mu_B)\ldots$$

This equation may be expanded, because eq. 7.10 links chemical potential to activities. Thus

$$\begin{aligned}\Delta G = \;& p\mu_P^0 + pRT\ln a_P \\ &+ q\mu_Q^0 + qRT\ln a_Q \ldots \\ &-(a\mu_A^0 + aRT\ln a_A) \\ &-(b\mu_B^0 + bRT\ln a_B)\ldots.\end{aligned} \quad (7.12)$$

The symmetrical arrangement helps us to recognize that the terms in the first column add up to ΔG^0 (see eq. 7.11). If now the numbers of moles, p, q, a and b are put inside the logarithms, the equation may be rewritten as

$$\boxed{\Delta G = \Delta G^0 + RT\ln\frac{a_P^p a_Q^q \ldots}{a_A^a a_B^b \ldots}}\,. \quad (7.13)$$

This is the **general reaction isotherm,** also known as the van't Hoff isotherm; it is of prime importance. The logarithmic ratio is sometimes known as the *activity quotient*, and is written Q. As before, ΔG is a measure of the affinity of the process actually occurring, where the logarithmic term makes adjustment for non-unit activities. This equation would apply for example when it was required to determine the feasibility of a reaction for which all starting activities are known.

If now the reaction is allowed to proceed, the activities will alter, and finally the reaction will reach equilibrium. At this point, ΔG is zero, and so eq. 7.13 becomes

$$\Delta G^0 = -RT\ln\frac{a_P^p a_Q^q \ldots}{a_A^a a_B^b \ldots}.$$

The activities now have an extra significance, because taken together they define the equilibrium condition, and are fixed among themselves. If T is fixed, then so is ΔG^0. This means that the activity term must also be constant; it is now equal to *the equilibrium constant* in terms of activities, K_a.

First critical comment Because activity has been carefully defined as a *dimensionless ratio*, K_a must also be dimensionless. For gases, with the standard state defined as unit fugacity, $a = f/f^0 \simeq p/p^0 = p$ atm/1 atm.

Finally the standard equilibrium isotherm may be written down:

$$\boxed{\Delta G^0 = -RT\ln K_a} \quad (7.14)$$

Despite its simplicity, this is probably the most widely useful equation in all of thermodynamics. It is now that we can appreciate the importance of ΔG^0 data; values for different reactions may be combined to give the equilibrium constant for *any other reaction of interest.*

Example We shall now apply the concepts of activity and equilibrium which we have developed to a particular reaction. A laboratory study has provided information on the reaction

$$FeO\,(l) + CO\,(g) \rightleftharpoons Fe\,(c) + CO_2\,(g),$$

over the temperature range 1693 K (the melting point of FeO), to 1808 K (that of Fe). The data is expressed linearly in temperature:

$$\Delta G^0\,(J/mol) = -50\,540 + 41.29T.$$

This reaction includes crystalline, liquid and gaseous phases, but even if solutions are present, the activities may be determined for each component. For simplicity, however, we shall assume that both condensed phases are pure and have unit activity.

By using this data, let us calculate the equilibrium constant at 1700 K. Firstly we find that ΔG^0_{1700} is +19.65 kJ/mol.

Second critical comment The standard states are those appropriate to a temperature of 1700 K. These are, for carbon monoxide and dioxide, the gases at unit fugacity, for iron the regular crystal stable at that temperature (alpha phase), and for ferrous oxide, the pure liquid.

Using eq. 7.14, we may write, for the equilibrium situation

$$\Delta G^0 = -RT \ln \frac{a_{Fe}a_{CO_2}}{a_{FeO}a_{CO}}$$

$$19\,650 = -8.314(1700)2.303 \log \frac{1 . p_{CO_2}(atm)/1\ atm}{1 . p_{CO}(atm)/1\ atm}.$$

Notice that because iron and ferrous oxide are assumed pure, their activities are put equal to one, and also that fugacity ratios for carbon monoxide and dioxide are put equal to pressure ratios at this high temperature.

The pressure ratio is now seen to be identical to K_p, the equilibrium constant expressed in terms of pressures.

(N.B. In some cases the equilibrium constant K_c is applied to gaseous systems. This term is not needed at this stage, and leads to complications; extra terms involving $(RT)^{\Delta n}$, where Δn is the increase in the number of gaseous moles when the reaction goes from left to right, are brought in when expressing K_c in terms of K_a or K_p. The concept of partial pressures is both simpler and more rewarding, and will be used in preference in this book.)

Finally we find that $K_a = p_{CO_2}/p_{CO} = 0.248$, so that carbon monoxide is only partly effective as a reducing agent at this temperature. So important is this kind of equilibrium that diagrams (named after Richardson) of equilibrium p_{CO_2}/p_{CO} ratios are plotted for many coke reduction processes over a wide range of temperature as an aid to process planning.

Summary We have derived a deceptively simple equation, $\Delta G^0 = -RT \ln K_a$, which enables us to calculate equilibrium constants from ΔG^0 data. Having derived and discussed the concept of activity, we can easily take into account the non-ideal behaviour of solutions, and of gases at significant pressures. However, in cases where solutions are dilute, or gas pressures low, calculations for the ideal case are also easily made, by putting all activity coefficients equal to unity.

PROBLEMS

7.1 At 2100 K, the equilibrium constant for the reaction

$$Si\ (l) + 2\ H_2\ (g) \rightleftharpoons SiH_4\ (g)$$

is $K = (0.996)10^{-6}$.

(a) What is ΔG^0 for this reaction?
(b) If ΔS^0_{2100} is -126.9 J/mol K, what is the enthalpy change?

7.2 An experimental study of cobaltous silicate, Co_2SiO_4, is launched. In a two pronged approach, the carbon monoxide reduction of Co_2SiO_4, and of CoO, are studied. For these reactions,

I $Co_2SiO_4\ (c) + 2\ CO \rightarrow 2\ CO_2 + 2\ Co\ (c) + SiO_2\ (c)$,
II $2\ CoO\ (c) \quad + 2\ CO \rightarrow 2\ CO_2 + 2\ Co\ (c)$

at 1300 K, the CO_2/CO pressure ratios were $(1.067)10^5$ for I, and $(1.835)10^5$ for II.

(a) Calculate ΔG^0_{1300} in each case.
(b) By combining these data suitably, find ΔG^0_{1300} for the combination of oxides,

$$2\ CoO + SiO_2 \rightarrow Co_2SiO_4.$$

7.3 For approximate determinations, ΔG^0_T may be calculated as $\Delta G^0_T = \Delta H^0_{298} - T\Delta S^0_{298}$. Use tabulated data to determine an equilibrium constant for one of the following industrial processes.

(a) *Manufacture of* CS_2

$$CH_4\ (g) + 2\ S_2\ (g) \xrightarrow[650°C]{Bauxite} 2\ H_2S\ (g) + CS_2\ (g).$$

(b) *Production of* HCN

$$CH_4 + NH_3 + 1\tfrac{1}{2}\ O_2 \xrightarrow[1000°C]{Pt/Rh} HCN + 3\ H_2O\ (g).$$

(c) *Acetylene, by the Sachse process*

$$\underset{\text{(natural gas)}}{2\ CH_4} \xrightarrow{1500°C} C_2H_2 + 3\ H_2 \ \text{(Rapid product cooling to avoid back reaction).}$$

(*Hint*: Use both $\Delta G^0_{f,298}$ and $\Delta H^0_{f,298}$ to find $\Delta S^0_{f,298}$, rather than deriving from individual S^0 data.)

7.4 In a study of the reaction

$$CO_2\ (g) + C\ (graphite) \rightleftharpoons 2\ CO\ (g)$$

at 796°C, it was found that at a total pressure of 1.046 atm the partial pressure of CO was 0.885 atm. Calculate the equilibrium constant for this reaction, and thence a value of ΔG^0 at this temperature.

7.5 Show how, using the third law of thermodynamics, the free energy change of a chemical reaction may be calculated from calorimetric data.

The standard free energy of formation of N_2O_4 at 25°C is 98.07 kJ/mol and that of NO_2 is 51.34 kJ/mol. Calculate the fraction of N_2O_4 that will be dissociated if a sample of the gas is allowed to attain equilibrium at 25°C and a total pressure of one atmosphere.

7.6 Tin of high purity is to be cast at a temperature of 600 K, at which ΔG^0 for the reaction

$$\text{Sn (l)} + \text{O}_2\text{ (g)} \rightarrow \text{SnO}_2\text{ (c)}$$

is -418.4 kJ/mol. Argon is used as a blanketing gas, but usually contains oxygen at a partial pressure of 10^{-6} atm. Will tin oxidize in this environment?

7.7 For the reaction: $ZnSO_4.7H_2O \rightarrow ZnSO_4.6H_2O + H_2O$ (l), $\Delta G = 1.477$ kJ/mol at 18°C. The vapour pressure of water is 15.48 mmHg at the same temperature. What is the vapour pressure of zinc sulphate heptahydrate in equilibrium with the hexahydrate at 18°C?

(*Hint*: the water vapour pressure gives the free energy change for the vaporization process.)

7.8 Use tabulated data to find the partial pressure of fluorine atoms at equilibrium in fluorine gas at 1 atm and at 298 K.

7.9 R. H. Newton and B. F. Dodge (*J. Am. chem. Soc.* **56,** 1287 (1934)) found that K_p for the reaction

$$\text{CO} + 2\,\text{H}_2 \rightleftharpoons \text{CH}_3\text{OH (g)}$$

was $(2.32)10^{-3}$ at 250°C. Calculate ΔG^0 for this reaction at this temperature.

Using tabulated data, find ΔG^0 at 298 K. Why is it necessary in practice to carry out this methanol synthesis at high temperature and pressure rather than at 298 K?

7.10 On heating, nitrosyl chloride NOCl decomposes into nitric oxide and chlorine:

$$2\,\text{NOCl} \rightarrow 2\,\text{NO} + \text{Cl}_2.$$

The reaction is endothermic, with $\Delta H^0 = 77.07$ kJ/mol Cl_2. If it may be assumed that both ΔH and ΔS are temperature invariant, and the standard molar entropies for NOCl, NO, and Cl_2 are 261.6, 210.7, and 223.0, respectively, calculate the equilibrium constant for the reaction at 650 K.

8

Equilibrium Experiments and Their Interpretation

In this chapter, we shall derive one further equation, the reaction isochore (sometimes called 'isobar'), and apply this and other equations to a variety of experiments. We shall see that relatively simple measurements of equilibria, taken over a range of temperature, provide information on both enthalpy changes, ΔH, and entropy changes ΔS, and therefore on changes of free energy ΔG.

8.1 The Reaction Isochore Equation

This equation is often linked with the name of van't Hoff, who was the first to apply it widely. It may be derived in several ways. We shall adopt the simplest method, which will make full use of the concept of activity. We start with the two equations 6.3 and 7.19, which should by now be engraved indelibly on the minds of all readers. They are

$$\Delta G = \Delta H - T\Delta S, \qquad (T \text{ constant}) \ (6.3)$$

which is, for standard conditions,

$$\Delta G^0 = \Delta H^0 - T\Delta S^0, \qquad (T \text{ constant}) \ (8.1)$$

and

$$\Delta G^0 = -RT \ln K_a. \qquad (7.19)$$

If ΔG^0 is eliminated, we obtain

$$\boxed{\ln K_a = -\frac{\Delta H^0}{RT} + \frac{\Delta S^0}{R}} \qquad (T \text{ constant}) \ (8.2)$$

This is the van't Hoff isochore (or isobar) equation. We shall use this equation often, although, at first sight, it seems not to be very helpful. Strictly speaking, there are *four* variables, K_a, ΔH^0, T, and ΔS^0, although only K_a varies widely with temperature. However, it is worth while to look more closely at ΔH^0 and ΔS^0, and their variation with temperature.

From eq. 3.4 we know that the variation of ΔH is given by

$$\frac{\mathrm{d}(\Delta H)}{\mathrm{d}T} = \Delta C_P.$$

When integrated, this equation states that

$$\Delta H_2 - \Delta H_1 = \int_{T_1}^{T_2} \Delta C_P \, \mathrm{d}T.$$

Thus the difference of heat capacities across the reaction determines the change of ΔH from T_1 to T_2. Usually, the heat capacities of the products almost equal those of the reactants, and ΔC_P is very small indeed. Often, ΔH is assigned a constant value. Over large ranges of temperature, the integral must be evaluated exactly; we shall see to this in a later chapter.

The argument for changes in ΔS runs on parallel lines. For a single material, because $dS = dq/T$,

$$\int_{T_1}^{T_2} dS = \int_{T_1}^{T_2} \frac{C_P\,dT}{T}$$

and so

$$S_2 - S_1 = \int_{T_1}^{T_2} \frac{C_P\,dT}{T}.$$

For a reaction, we are concerned with ΔC_P, and so

$$\Delta S_2 - \Delta S_1 = \int_{T_1}^{T_2} \frac{\Delta C_P\,dT}{T}.$$

As before, ΔC_P is often very small, and this equation need be calculated only for large changes of temperature. (Note that although ΔS^0 may be temperature-invariant, absolute entropies are certainly not.)

Let us return to the isochore equation (eq. 8.2). To recapitulate: we have shown that ΔH^0 and ΔS^0 may be regarded as constant over small ranges of temperature, and so the equation will simply show the variation of K_a with temperature. Let us therefore modify it, and write it simply as

$$\ln K_a = -\frac{\Delta H^0}{R}\cdot\frac{1}{T} + C \qquad (T \text{ constant}) \quad (8.3)$$

where C is a constant, equal to $\Delta S^0/R$. Notice that the restriction on constant temperature applies only to a particular experiment, which lasts only long enough to take the necessary measurements. Further experiments at different temperatures would be quite possible. We should find that K_a would vary exponentially with $1/T$ if eq. 8.3 is valid. Further, if we recognize that eq. 8.3 has the classic form of the $y = mx + c$ relationship, we see that y corresponds to $\ln K_a$, x to $1/T$, and m, the gradient, to $-\Delta H^0/R$; that is, the slope of the straight-line graph,

$$\frac{d(\ln K)}{d(1/T)} = -\frac{\Delta H^0}{R}.$$

We shall use this relationship later in the chapter.

It is possible, if data are scarce, to obtain meaningful results with only two data, preferably spaced well apart. The eq. 8.3 must hold for both sets of values 1 and 2, and so we can write

$$\ln K_1 = -\frac{\Delta H^0}{R}\cdot\frac{1}{T_1} + C,$$

and

$$\ln K_2 = -\frac{\Delta H^0}{R}\cdot\frac{1}{T_2} + C.$$

On subtracting these, we have

$$\ln\frac{K_1}{K_2} = -\frac{\Delta H^0}{R}\left(\frac{1}{T_1} - \frac{1}{T_2}\right). \quad (8.4)$$

This is a useful form of the equation for purposes of simple calculation.

One final point should be made concerning equilibria in condensed phases (no gases present). In such a case, K_a would be expressed in terms of concentrations or mole fractions, and we would be justified in putting K_c in its place. In addition, ΔH need no longer be used; ΔU may be substituted, because both pressure *and volume* will remain constant. A very simple proof would look like this:

because $\qquad H = U + PV,$

therefore $\qquad \Delta H = \Delta U + (P_2V_2 - P_1V_1),$

and because P and V are both constant (condensed phases),

$$\Delta H = \Delta U.$$

The isochore will now wear a very thin disguise, but it will be recognizable none the less, as

$$\ln K_c = -\frac{\Delta U^0}{RT} + \frac{\Delta S^0}{R}. \qquad (P,\ V \text{ constant})$$

8.1.1 Le Chatelier up to Date

Because K is taken to be a function of temperature only, we may differentiate $\ln K$ with respect to T. Thus, if we write the isochore as

$$\ln K = -\frac{\Delta H^0}{R}.T^{-1} + C,$$

$$\boxed{\frac{\mathrm{d}(\ln K)}{\mathrm{d}T} = \frac{\Delta H^0}{RT^2}} \qquad (8.5)$$

Let us consider a particular reaction. Iodine vapour dissociates into atoms, to a progressively greater extent as the temperature is raised. Is this fact amenable to thermodynamic treatment? This equation shows us that it is. Looking at it qualitatively, we see that the sign of the right-hand side is the same as that of ΔH. For the reaction

$$I_2\,(g) \rightleftharpoons 2\,I\,(g), \quad \Delta H = +120.95 \text{ kJ/mol},$$

and so $\Delta H/RT^2$ is positive. This means that $\mathrm{d}(\ln K)/\mathrm{d}T$ is also positive, and so K increases logarithmically as T increases. Not only can we see this qualitatively, but if we are given K at one temperature, we can also make quantitative estimates of K at other temperatures. Thus, by using eq. 8.5, we see that at 300 K,

$$\mathrm{d}(\ln K)/\mathrm{d}T = 120\,950/8.314(300)^2 = 0.16.$$

That is, for every *ten* degrees, $\ln K$ changes by 1.6. This means that K increases by *five times every ten degrees*, in the region of 300 K. This change seems quite sensational but such effects are not uncommon.

This calculation is seen to be in the nature of a quantitative Le Chatelier prediction. He was able to predict the direction of changes in equilibrium with changes of temperature ('such that the effect of the change of conditions shall be minimized'), but we are now able to measure its magnitude. This we shall do in the examples which follow, where we shall use the integrated form of the isochore, given as eq. 8.2. The examples demonstrate different experimental techniques, and cover different types of process. *The basic thermodynamic data so obtained can be transferred and modified in order to predict and understand new reactions, and new processes.*

8.2 Applications of the Isochore Equation

We shall discuss three equilibrium studies. New aspects of the isochore will be brought out as they arise.

8.2.1 Vaporization Processes

The equilibrium vapour pressure of liquid gallium over a range of temperature has been measured and is summarized in Table 8.1.

The Knudsen effusion technique entails heating a small quantity of the material under study in a refractory container, in a high vacuum. Vapour effuses through a small pinhole in the lid, and sprays on to a cold surface, where it condenses. After several hours, the deposit is weighed. Vapour pressures, such as those in Table 8.1, may then be calculated from the pinhole size, the deposition time, the weight deposited, and other constant factors. If the pinhole is sufficiently small, then equilibrium conditions may be assumed to prevail inside the cell.

Table 8.1 *Vapour pressure of liquid gallium in the Knudsen effusion technique, from 1281 to 1437 K*[a]

Temperature, K	Vapour pressure, atm	$1/T$ K	log p (atm)
1281	$(7.00)10^{-6}$	$(7.81)10^{-4}$	−5.155
1311	$(1.19)10^{-5}$	$(7.63)10^{-4}$	−4.924
1330	$(1.79)10^{-5}$	$(7.52)10^{-4}$	−4.747
1355	$(2.60)10^{-5}$	$(7.38)10^{-4}$	−4.585
1368	$(2.82)10^{-5}$	$(7.31)10^{-4}$	−4.550
1373	$(3.59)10^{-5}$	$(7.28)10^{-4}$	−4.445
1390	$(4.36)10^{-5}$	$(7.19)10^{-4}$	−4.360
1413	$(7.45)10^{-5}$	$(7.08)10^{-4}$	−4.128
1437	$(1.00)10^{-4}$	$(6.96)10^{-4}$	−4.000

[a] Alcock, C. B., Cornish, J. B. and Grieveson, P., *Proc. Conf. on Thermodynamics*, International Atomic Energy Authority, Vienna (1966).

The equilibrium process may be represented as

$$\mathrm{Ga\,(l)} \rightleftharpoons \mathrm{Ga\,(g)}$$

$$a = 1 \qquad f \approx p.$$

The activity of the liquid is taken as unity, and for the vapour it is valid to assume that fugacity and vapour pressure are equal. Then the equilibrium constant K_a, which is $a_{\mathrm{Ga(g)}}/a_{\mathrm{Ga(l)}}$, becomes equal to p, the vapour pressure. Moreover, the enthalpy change is ΔH_{vap} for the vaporization process. Hence eq. 8.2 becomes

$$\ln p\,(\mathrm{atm}) = -\frac{\Delta H_{\mathrm{vap}}}{R}\cdot\frac{1}{T} + C.$$

If we use decadic logarithms, we have

$$\log p\,(\mathrm{atm}) = -\frac{\Delta H_{\mathrm{vap}}}{2.303R}\cdot\frac{1}{T} + \frac{C}{2.303}. \qquad (8.6)$$

Values of $\log p$ and $1/T$ have been calculated, and also appear in Table 8.1. Figure 8.1 shows a graph of $\log p$ versus $1/T$, which gives a good straight line.

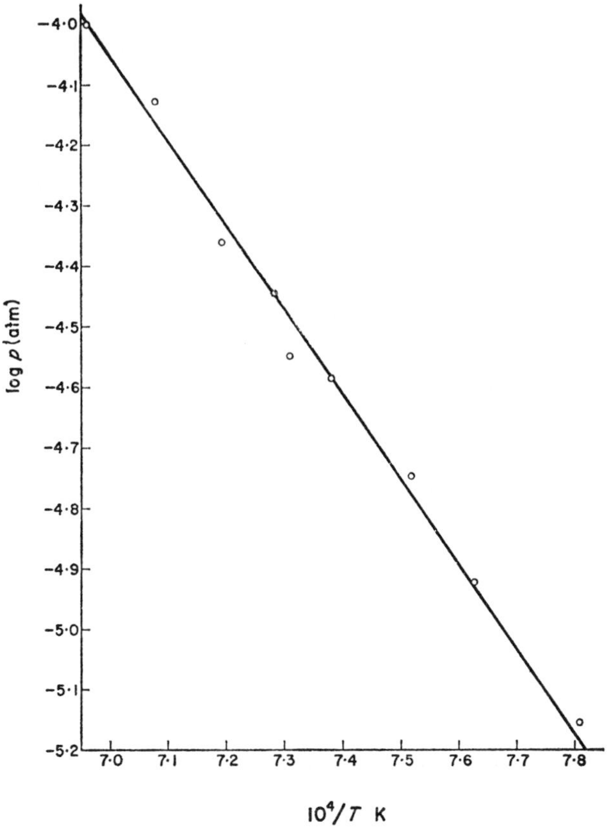

FIG. 8.1 Vapour pressure of gallium, expressed as a $\log p$ (atm) versus $1/T$ graph, from 1280 to 1440 K.

The gradient is near −14 040 (K), and so

$$\Delta H_{\text{vap}} = -2.303R(-14\,040) = \mathbf{+268.8\ kJ/mol.}$$

When sample values of p and T are put into the equation, we find the constant term $C/2.303$ to be 5.777. We previously put $C = \Delta S/R$, and so

$$\Delta S = CR = 2.303R(5.777) = \mathbf{110.5\ J/mol\ K.}$$

This value is the entropy change on converting one mole of liquid gallium at about 1400 K to vapour at the same temperature, and a pressure of *c.* 5×10^{-5} atm. Strictly speaking, because ΔH will have a slightly different value at the boiling point, one should be cautious when comparing this value with Trouton's constant. None the less, the agreement between the indicated value 110.5, and values for mercury (94.1), cadmium (103.8), and zinc (105.4) is encouraging. As a final check on the vapour pressure equation as it now stands,

$$\log p\ (\text{atm}) = -\frac{268\,800}{2.303R}\cdot\frac{1}{T} + 5.777,$$

we can calculate the boiling temperature, where $p = 1$ atm. It is

$$T_b = 268\,800/2.303R(5.777) = \mathbf{2430\ K.}$$

Quoted values of the boiling point vary from 2256 to 2344 K.

The vapour pressures of most liquids conform to the simple relationships discussed above. To demonstrate that this is so for high boiling metals as well as for lower boiling organic materials, a number of plots of log (vapour pressure) versus $1/T$ are shown in Fig. 8.2. They are clearly members of a family of lines, which seem to converge at a point where $1/T = 0$. Although this has little physical

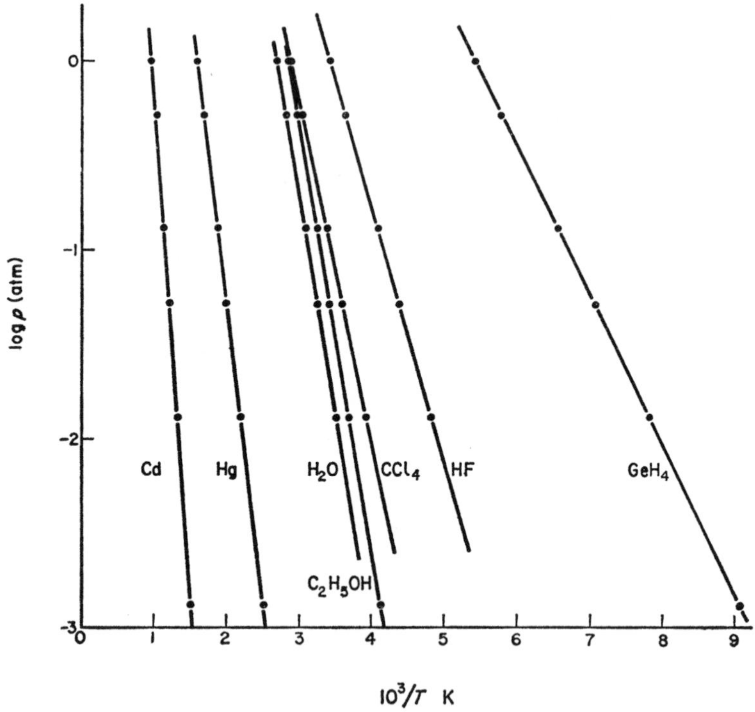

FIG. 8.2 Graphs of log p versus $1/T$ for the vaporization of a variety of materials.

meaning, it can be interpreted algebraically as the point where $C = 2.303 \log K = \Delta S/R$. From such a series of graphs, C is found to vary from about 10 to 15, corresponding to values of ΔS of between 80–125 J/mol K. This wide variation only serves to underline the fact that Trouton's rule is at best a rule of thumb.

8.2.2 THE DECOMPOSITION OF THE COMPOUND UIn_3

Uranium forms a stable tri-indide. When heated at over 1000 K, it begins to decompose, according to the equation

$$UIn_3\,(s) \rightleftharpoons U\,(\gamma) + 3\,In\,(g). \tag{8.7}$$

At these high temperatures, uranium exists as the gamma crystalline phase, as opposed to the alpha, or room temperature form. If a mixture of UIn_3 and uranium is heated, we can be sure that no free liquid indium will be present.

However, it is still possible to measure an indium vapour pressure, which now takes on a new and important role. Unlike our first example, there is no simple vaporization process occurring here. Now, the indium vapour pressure can be seen as the 'pulse' of the decomposition reaction. The equilibrium constant is

$$K_a = \frac{a_U \times a_{In}^3}{a_{UIn_3}} = p_{In}^3.$$

We can therefore derive much thermodynamic information from a series of vapour pressure measurements. Thus, at 1250 K, $p_{In} = (9.49)10^{-3}$ mmHg or $(1.25)10^{-5}$ atm. Therefore,

$$\Delta G^0_{1250} = -RT \ln K = -2.303(8.314)(1250)(3) \log (1.25)10^{-5}$$
$$= +\ \mathbf{352.1\ kJ/mol.}$$

This measurement, and others over a range of temperatures from 1100–1300 K have been taken using a Knudsen cell effusion technique, and are summarized in Table 8.2.

TABLE 8.2 *Equilibrium constants of the* UIn_3 *decomposition reaction, from vapour pressure measurements*

Temperature, K	p_{In}, atm	K_a, as for eq. 8.7
1126	$(6.07)10^{-7}$	$(2.24)10^{-19}$
1156	$(1.34)10^{-6}$	$(2.41)10^{-18}$
1230	$(7.98)10^{-6}$	$(5.08)10^{-16}$
1256	$(1.43)10^{-5}$	$(2.92)10^{-15}$
1295	$(3.27)10^{-5}$	$(3.50)10^{-14}$

The isochore equation may now be solved, graphically or otherwise, to give values of

$$\Delta H^0 = \mathbf{857.3\ kJ/mol}$$

and

$$\Delta S^0 = \mathbf{404.2\ J/mol\ K,}$$

that is

$$\ln K_a = \frac{-857\,300}{RT} + \frac{404.2}{R}.$$

We now have sufficient data to express the Gibbs free energy change of the reaction as

$$\Delta G^0 = 857\,300 - 404.2T.$$

Assuming that $\Delta C_P = 0$, this equation may be used to find the standard free energy of formation of UIn_3. At 25 °C, it would be produced from *alpha* uranium and *liquid* indium. Both the phase change of uranium, and the vaporization of indium have been studied, and the following relationships determined;

(a) $\quad In\ (l) \rightleftharpoons In\ (g), \quad \Delta G^0 = +251.2 - 116.4T,$

(b) $\quad U\ (\gamma) \rightleftharpoons U\ (\alpha), \quad \Delta G^0 = -7740 + 7.74T.$

On combining these with the equation

(c) $\quad UIn_3\ (c) \rightleftharpoons U\ (\gamma) + 3In\ (g), \quad \Delta G^0 = 857\,300 - 404.2T,$

we obtain the equations for formation of UIn_3. They are

$$U\,(\alpha) + 3In\,(l) \longrightarrow UIn_3\,(c), \quad \Delta G^0 = -96\,150 + 62.76T.$$

This gives a value for the standard free energy of formation at 25°C of

$$\Delta G^0_{298} = -96\,150 + 62.76(298)$$
$$= \mathbf{-77.45\ kJ/mol.}$$

8.2.3 The High Temperature Dissociation of Water Vapour

As a final example, we shall consider the origins of the data on that most important of species, the water molecule. Such quantities as ΔG^0 and ΔH^0 for the formation of water are used so frequently that it is essential to have reliable, accurate data to draw upon. For example, water dissociation equilibria will be an important factor in flame and rocket technology. Such processes are strongly endothermic, and will run away with the available energy. This will limit the maximum temperature of the flame, as pointed out in Chapter 3.

In this example, the equilibrium data depend on determining not one but three separate partial pressures, and more sophisticated tools, such as ultra-violet absorption and emission spectrometers were used to determine the equilibrium constant K at different temperatures. One of the reactions studied was

$$2\,H_2 + O_2 \rightleftharpoons 2\,H_2O, \quad \text{for which}$$

$$K_a = \frac{a_{H_2O}{}^2}{a_{H_2}{}^2 \times a_{O_2}} = \frac{p_{H_2O}{}^2}{p_{H_2}^2 \times p_{O_2}}.$$

The pressures were low at all times, and so $\gamma = 1$ throughout. The dissociation is very slight, and so the water vapour pressure is virtually equal to the total vapour pressure. In Table 8.3 are a few of the values of K calculated from such experimental results.

It is most striking to see such vast changes occur in the equilibrium constant over a few hundred degrees. For example, from the temperature of boiling water to that of molten lead, 327°C, K increases by a factor of over 10^{25}. However, even such changes as these are amenable to the isochore equation. It is suggested that the student should draw his own graph to verify the results of the calculations which follow. From the gradient and a spot value on the graph of log K versus $1/T$ K, we obtain the equation

$$\log K = +\frac{25\,510}{T} - 5.25.$$

This quickly yields values for

$$\Delta H^0 = -2.303(8.314)(25\,510) = \mathbf{-488.3\ kJ/mol,}$$

and $$\Delta S^0 = 2.303(8.314)(-5.25) = \mathbf{-100.5\ J/mol\ K.}$$

As the quoted reaction is twice the equation for the formation of water vapour, which would be

$$H_2\,(g) + \tfrac{1}{2}\,O_2\,(g) \rightleftharpoons H_2O\,(g),$$

a value for $\Delta H^0_{f,298}$ of $-488.3/2 =$ **−244.1 kJ/mol** is given. This is close to the tabulated value in Chapter 2. The standard free energy change for the reaction is easily found.

$$\Delta G^0 = \Delta H^0 - T\Delta S^0 = -488.3 - 298.15\,(-100.5)/1000$$
$$= -458.3\ \text{kJ/mol}.$$

From this, $\Delta G^0_{f,298}(H_2O) = -458.3/2 = \mathbf{-229.1\ kJ/mol.}$

This compares well with the currently accepted value of -228.6 kJ/mol.

TABLE 8.3 *Equilibrium constants at various temperatures for the formation of two moles of water vapour from hydrogen and oxygen*

T K	K	$\log_{10} K$
400	$(3.4)10^{58}$	58.53
500	$(5.25)10^{45}$	45.72
600	$(2.0)10^{37}$	37.30
800	$(4.0)10^{26}$	26.60
1000	$(1.4)10^{20}$	20.14
1200	$(6.6)10^{15}$	15.82
1500	$(2.9)10^{11}$	11.46
2000	$(1.25)10^{7}$	7.10

8.3 The Clapeyron Equation

Section 8.2.1 was concerned with equilibrium between a condensed phase and the vapour. It is often necessary, however, to estimate the effect of pressure on equilibria between *two condensed phases.* For example, the melting point of sodium at one atmosphere pressure is 97.6°C. Can it be used as a liquid heat transfer medium at 100°C, at a pressure of 100 atm, or will it solidify? It is known that the liquid is less dense than the solid, and this argues that high pressures will encourage solidification. This is another aspect of Le Chatelier's work, which we can now quantify. This and similar problems may be solved by the Clapeyron equation, which we shall now derive.

Let us continue with the sodium example, representing the process as Na (s) $\rightleftharpoons$ Na (l). Because solid and liquid sodium are in equilibrium at 97.6°C, and 1 atm, their chemical potentials must be equal, i.e., $\mu_s = \mu_l$. Suppose that we can now arrange a different pressure and temperature which allow equilibrium to occur once again. Although the chemical potentials will have changed, they will have done so *by the same amount*; this is necessary to maintain the new potentials equal, that is $\mu'_s = \mu'_l$. Therefore it follows that $d\mu_s = d\mu_l$. We shall now use eq. 7.1, expressed in molar quantities (usually indicated by the bar). Applied to the solid, it tells us that $d\mu_s = \bar{V}_s\,dP - \bar{S}_s\,dT$.

Also, for the liquid, $d\mu_l = \bar{V}_l\,dP - \bar{S}_l\,dT$.

Because

$$d\mu_s = d\mu_l,$$

$$\bar{V}_s\,dP - \bar{S}_s\,dT = \bar{V}_l\,dP - \bar{S}_l\,dT$$

or

$$\frac{dP}{dT} = \frac{\Delta\bar{S}}{\Delta\bar{V}}.$$

Finally, our equilibrium situation requires that $\Delta G = \Delta H - T\Delta S = 0$, and so $\Delta S = \Delta H/T$. We can now write the Clapeyron equation as

$$\boxed{\frac{dP}{dT} = \frac{\Delta H}{T\Delta\bar{V}}} \qquad (8.8)$$

The sodium problem may now be tackled. We find the following data:

$$\text{Heat of fusion} = 3.05 \text{ kJ/mol},$$
$$\text{Density of liquid} = 0.928\,8 \text{ g/ml},$$
$$\text{Density of solid} = 0.951\,9 \text{ g/ml}.$$

Thus

$$\Delta H_{fus} = 30.1 \text{ litre atm/mol},$$
$$(1 \text{ joule} = 0.00987 \text{ litre atm})$$
$$\Delta \bar{V} = +6.05 \times 10^{-4} \text{ litre/mol}.$$

From eq. 8.8 we calculate

$$\frac{dP}{dT} = \frac{30.1}{370.8(6.05 \times 10^{-4})} = 134 \text{ atm/K}.$$

A one degree rise in melting point to 97.6 + 1 = 98.6°C is accompanied by an increase of pressure of 134 atm, and so use of the liquid sodium at the suggested 100 atm and 100°C would give no problems. Liquid sodium is in fact used as a coolant in some types of nuclear reactor.

Summary We have seen how data from experimental studies of equilibria are converted into a readily usable form, that is, into ΔH, ΔS, and ΔG values. These data are combined with others, determined, for example, by reaction calorimetry (ΔH) or by 'third law' (as opposed to 'second law') determinations of entropy, using low temperature heat capacity measurements. Before we can enter the first division of predictive thermodynamics, and confidently plan processes, we must learn to accept information from two further sources. First, there is a great reservoir of electrochemical expertise which we have not yet tapped. Secondly, we must learn to handle the refinements in free energy formulations, which make allowance for the slight variations of ΔH and ΔS with temperature.

PROBLEMS

8.1 B. H. Zimm and J. E. Mayer (*J. chem. Phys.* **12,** 362 (1944)) accurately determined the vapour pressure of solid KBr, over a range of temperature. Determine $\Delta H_{sublimation}$ from their data graphically.

T K	P, atm	T K	P, atm
900	$(3.93)10^{-5}$	746	$(1.045)10^{-7}$
877	$(1.55)10^{-5}$	720	$(3.22)10^{-8}$
835	$(4.25)10^{-6}$	696	$(9.54)10^{-9}$
813	$(1.855)10^{-6}$	680	$(4.16)10^{-9}$
777	$(4.25)10^{-7}$	658	$(1.20)10^{-9}$
757.5	$(1.95)10^{-7}$	637	$(3.38)10^{-10}$

8.2 The vapour pressure of water at 25°C is 23.76 mmHg. Use this datum and the normal boiling point of water to derive an equation for the temperature variation of the vapour pressure of water from 25°–100°C. A mountaineer encounters an atmospheric pressure of only 485 mmHg; at what temperature will water boil?

8.3 The vapour pressure of solid and liquid Tc_2O_7 is as follows:

Solid		Liquid	
T K	P(mmHg)	T K	P(mmHg)
391.2	0.65	393.3	1.0
388.7	0.46	413.1	2.1
384.1	0.36	427.5	4.0
377.9	0.18	446.4	9.6
373.2	0.095	459.8	16.7
368.5	0.05	473.8	29.0
362.2	0.03	489.7	54.6

Determine the enthalpies of evaporation and sublimation over the temperature range of study. What is the melting point?

8.4 The equilibrium constant K_p for the dissociation of bromine vapour, $Br_2 \rightleftharpoons 2Br$ changes with temperature as follows:

T K	K_p
600	$(6.18)10^{-12}$
800	$(1.02)10^{-7}$
1000	$(3.58)10^{-5}$
1400	$(3.03)10^{-2}$
1600	$(2.55)10^{-1}$

Determine ΔH_f^0 of bromine atoms over this temperature range, given that $\Delta H^0{}_{\text{vaporization}} = 30.92$ kJ/mol Br_2.

8.5 A study of the equilibrium: $CO_2 + C \rightleftharpoons 2\,CO$ yielded the following results:

T K	Total pressure (atm)	Mole % CO_2 in equilbrium mixture
1073	2.57	26.45
1173	2.30	6.92

For the equilibrium: $2\,CO_2 \rightleftharpoons 2\,CO + O_2$ the value of K_p at 1173 K is $(1.25)10^{-16}$ atm; the heat of combustion of carbon to carbon dioxide at 1173 K is 390.7 kJ/mol. Calculate ΔH and ΔS for the reaction:

$$2\,CO_2 \rightleftharpoons 2\,CO + O_2 \quad \text{at} \quad 1173\text{ K}.$$

8.6 At 251 °C, K_p for the dissociation of PCl_5

$$PCl_5 \rightleftharpoons PCl_3 + Cl_2$$

is found to be 2.684. At 307°C, it has reached 25.45. Calculate ΔG^0 at each temperature, and also the average value of ΔH^0 for this temperature range. Compare this data with current values in Table 2.1.

8.7 If oxygen gas is irradiated, O_2^+ species are formed, which rapidly equilibrate with oxygen: $O_2^+ + O_2 \rightleftharpoons O_4^+$. Using a mass spectrometer, J. Yang and D. C. Conway (*J. chem. Phys.* **40,** 1729 (1964)) found the ratio of O_4^+/O_2^+ at various temperatures by measuring peak-heights. Oxygen pressure was taken to be the total pressure, 1 mmHg, and values of K were calculated:

T K	259	273	293	320	344
K_p	2890	990	312	73.7	27.4.

Determine graphically ΔH for the process (this gives the bond dissociation energy for the $(O_2 - O_2^+)$ bond).

8.8 Boron nitride is a refractory compound of great strength which is finding application as a fibre material. For its formation by the reaction:

$$B\,(c) + \tfrac{1}{2}\,N_2\,(g) \longrightarrow BN\,(c),$$

K_p is 173.15 atm$^{-\frac{1}{2}}$ at 1900 K, but it drops to 38.94 atm$^{-\frac{1}{2}}$ at 2100 K. What is the enthalpy change for formation at 2000 K?

This datum may be used to calculate the bond dissociation energy of (B—N) in the crystal, which is of diamond structure. (How many moles of bonds in a mole of BN?) For 2000 K, use the following heats of atomization:

$$\Delta H^0_{f,2000}(B\,(g)) = 556\ \text{kJ/mol},$$
$$\Delta H^0_{f,2000}(N\,(g)) = 452\ \text{kJ/mol}.$$

8.9 The ionic product of water, $k_w = [H^+][OH^-]$varies with temperature. It is only $(0.2920)10^{-14}$ at 10.0°C, but has risen to $(2.919)10^{-14}$ at 40.0°C. Use this information to calculate the enthalpy change of neutralization of a strong acid by a strong base, both being of unit activity. Using this value and standard data, calculate $\Delta H^0_{f,298}(OH^-(aq))$.

8.10 Sulphur melts at 119.0°C under 1 atm, and $\Delta H_{\text{fusion}} = 55.2$ J/g. Densities are 1.811 g/ml (liquid), and 2.050 g/ml (solid). In the Frasch process, molten sulphur is pumped from underground deposits at 6 atm total pressure. What must be its minimum temperature? (1 J = $(9.87)10^{-3}$ l atm.)

8.11 Aluminium expands slightly on melting, and $\Delta V = +(1.9)10^{-5}$ l/g. The melting point is 658°C, and $\Delta H_{\text{fusion}} = 393$ J/g. A sensitive probe is being developed to determine very high pressures, by measuring the melting temperature. A change of 0.001 °C can be detected. What change of pressure does this represent? (1 J = $(9.87)10^{-3}$ l atm.)

9

Electrochemistry

'Her own mother lived the latter years of her life in the horrible suspicion that electricity was dripping invisibly all over the house.'

(James Thurber, *My Life and Hard Times*)

In fact, James Thurber's grandmother was partly right; many of the chemical processes occurring around us involve the movement of charged ions in liquids, as well as of electrons in metals. Many corrosion processes arise when aqueous electrolytes are in contact with iron, steel, or copper, and battery-powered appliances are common.

Electrochemical cells give valuable thermodynamic information, as well as being useful in their own right. Measurements of cell e.m.f. yield free energy changes directly, and tabulations of electrode potentials provide an alternative source of free energy data. Variations of e.m.f. with concentration lead to a better understanding of activities, and we shall also see how to measure entropy changes from the variation of cell e.m.f. with temperature.

9.1 Electrochemical Cells

A cell is a device for harnessing the energy of a chemical process to do electrical work, *directly*. A cell **electromotive force** (e.m.f.) is available at the **electrodes,** which are immersed in, or in contact with, the **electrolyte,** in which charged species, **ions,** are mobile.

If a bar of zinc is dipped into a solution of zinc sulphate, some zinc ions, Zn^{2+},

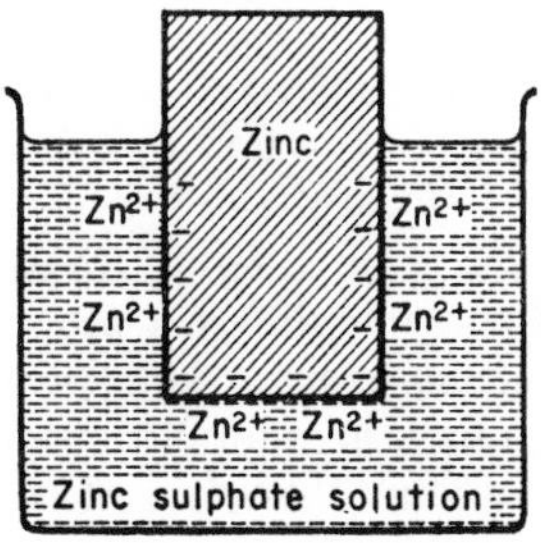

FIG. 9.1 The ions continue to leave the metal as long as they can overcome the opposing electrostatic forces. Electrons on the zinc attract them, the other zinc ions repel them.

dissolve, leaving two electrons each on the metal. This causes a separation of charge, and eventually (Fig. 9.1) equilibrium is achieved. An electrical double layer forms, which consists of electrons on the metal surface, and zinc ions immediately adjacent to it. At this stage, the tendency to dissolve is exactly matched by the tendency of zinc ions to deposit, which is caused by the charge separation. This means that there is a *potential difference between metal and solution*, which, however, *cannot be measured.* If the p.d. could be measured, we would have a direct measure of the free energy of formation of hydrated zinc ions,

$$Zn\ (c) \longrightarrow Zn^{2+}\ (aq) + 2e.$$

However, the second lead to the voltmeter would come from the solution, and this would constitute a second electrode. Moreover, reactions of this kind, which entail charge separation, occur to only a minute extent.

If we can construct a **cell** from two **half-cells** such as that described above, the e.m.f. does give free energy information as long as certain rules are obeyed. As an example, consider the Daniell cell, in which a porous barrier separates zinc in zinc sulphate solution from copper in copper sulphate solution. Such a cell may be depicted by the **cell-diagram**

$$^{\ominus}Zn \mid ZnSO_4(aq) \parallel CuSO_4(aq) \mid Cu^{\oplus},$$

where a single line represents the interface between an electrolyte and another phase, and the double line represents the link between the two solutions. The reaction equation, *by convention*, is written *starting from the left*, that is, with the Zn electrode:

$$Zn + Cu^{2+}\ (aq) = Zn^{2+}\ (aq) + Cu.$$

The signs at the electrodes represent the charge accumulating at each. Copper is the more noble metal of the two, and cupric ions tend to deposit, so making the copper electrode positive. No reaction occurs until these charges are dispersed, usually by means of some outside circuit, in which electrons will flow from the zinc to the copper. The electrode reactions will then be:

$$Zn \longrightarrow Zn^{2+} + 2e, \text{ and } Cu^{2+} + 2e \longrightarrow Cu.$$

We could measure the cell e.m.f. (E) by matching it with an exactly equal, but opposite e.m.f. from a potentiometer. The cell e.m.f. will be a measure of the free energy for the cell reaction if the electrodes are truly *reversible* and if potentials at the liquid junction are negligible. Reversibility implies that an external e.m.f. infinitesimally greater than E causes the cell reaction to be reversed (cf. '*external pressure*' *for gaseous work*). This is not true of all cells.

In order to maintain electrical neutrality in each cell compartment, negative ions move from right to left, within the cell, and positive ions in the opposite direction. In general, they will have different transport numbers; that is, they move at different rates and so carry different proportions of the total current. This results in a small *liquid junction potential*, which disturbs the cell e.m.f. reading. (Space allows only a brief account of transport numbers to be included here. For a more complete account the works of Moore [1] and Potter [2] are recommended.) If, instead of using a porous plate, as in the simple Daniell cell, a **salt bridge** is used to link the two electrolytes, the junction potential is minimized. A salt bridge is generally a tube of agar jelly containing a solution of either KCl or NH_4NO_3, for which transport numbers of anion and cation are equal. Such a cell is represented as

$$^{\ominus}Zn \mid ZnSO_4\ (aq) \mid KCl \text{ bridge} \mid CuSO_4\ (aq) \mid Cu^{\oplus}.$$

This cell and the next to be discussed, are shown in Fig. 9.2.

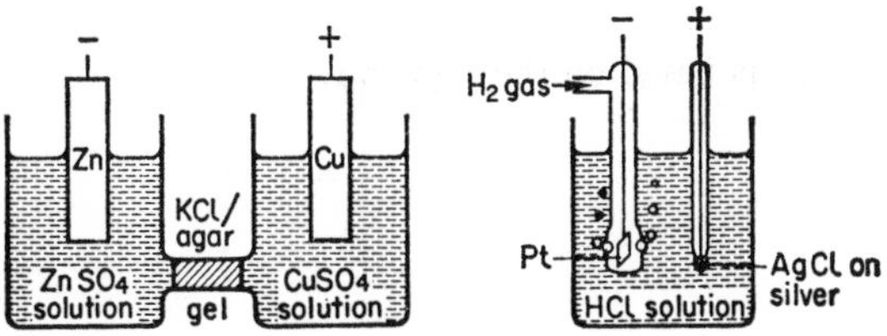

FIG. 9.2 Two cells discussed in the text. They exemplify the simple metal/ion electrode, the gas/ion electrode, and the metal/metal halide/ion electrode.

The second cell in Fig. 9.2 may be represented by

$$^{\ominus}\text{Pt, H}_2\text{ (gas)} \mid \text{HCl (aq)} \mid \text{AgCl, Ag}^{\oplus}$$

The hydrogen gas electrode comprises a platinized platinum foil with pure hydrogen bubbling over it. If the gas and the acid are at unit activity, then it is known as the *standard hydrogen electrode* (SHE). The silver-silver chloride electrode on the other hand operates as a source of chloride ions. The cell reaction and component half cell reactions are:

$$\begin{array}{lllll} \text{Left:} & \frac{1}{2}\,H_2 & \rightleftharpoons H^+ & + e \\ \text{Right:} & AgCl + e & \rightleftharpoons Cl^- & + Ag \\ \hline & \frac{1}{2}\,H_2 + AgCl & \rightleftharpoons H^+Cl^- & + Ag. \end{array}$$

Notice that hydrogen appears first, both in the cell diagram and in the cell equation. In this cell, liquid junction potentials do not arise, and if the electrodes are well prepared, reversibility is realized.

The zinc/zinc ion, the hydrogen, and the silver/silver chloride electrodes are typical of three common types of aqueous half-cell. A fourth type is the so-called redox half cell, which involves for example both ferrous and ferric ions in solution. A non-reacting electrode, usually of platinum, facilitates the following half-cell process:

$$Fe^{2+} \rightleftharpoons Fe^{3+} + \underset{\text{(from Pt)}}{e}$$

The half cell would be represented as Pt | Fe^{2+}, Fe^{3+}; it is basically similar to M | M^+ (M = metal), where Fe^{3+} is M^+, and Fe^{2+} is M.

9.2 Cell Energetics

If the reaction

$$\text{Zn (c)} + \text{CuSO}_4\text{ (aq, 0.5M)} \longrightarrow \text{ZnSO}_4\text{ (aq, 0.5M)} + \text{Cu (c)}$$

is performed in a calorimeter at 15°C, ΔH is found to be -235.2 kJ/mol. No useful work is done, and the process is quite irreversible.

Imagine, on the other hand, that the cell

$$^{\ominus}\text{Zn} \mid \text{ZnSO}_4\text{ (0.5M)} \mid \text{salt bridge} \mid \text{CuSO}_4\text{ (0.5M)} \mid \text{Cu}^{\oplus}$$

achieves the same overall change at 15°C, *but reversibly*. Maximum work is done when the load approaches infinite resistance, because the cell e.m.f. and the back e.m.f. are then infinitesimally different. This work is measured conveniently by immersing the resistor in a calorimeter. In addition, the cell itself is in another calorimeter, in order to measure q_{rev} of the cell.

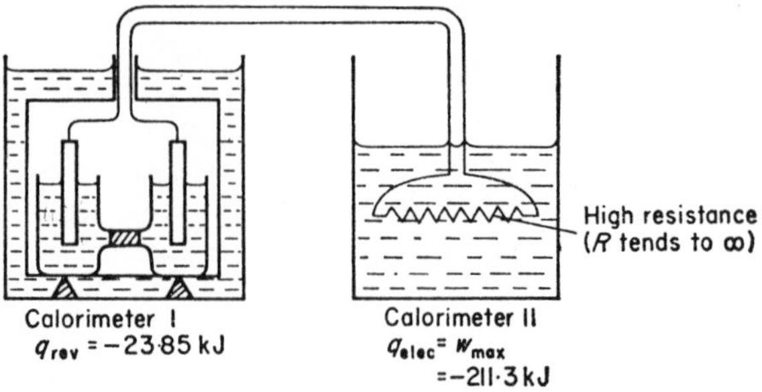

FIG. 9.3 In this experiment, the enthalpy of reaction is divided into that available to do work, measured as heat in calorimeter II, and that not available, even during reversible operation (calorimeter I).

Figure 9.3 summarizes the situation. The results show that most, but not all, of the enthalpy change is available as electrical energy, but 23.85 kilojoules are lost to us. Because the operation is reversible, we are already doing our best to maximize the useful work. Also because $q_{rev}/T = \Delta S$, the term $T\Delta S$ is -23.85 kJ, and $\Delta S = -23\,850/288 = -82.8$ joule/mol K.

The overall change of enthalpy is the same for each experiment, so

$$\begin{aligned}\Delta H &= q_{rev} + w_{max}\\ &= T\Delta S + w_{max}.\end{aligned}$$

(Note that w_{max} represents the work other than that due to gaseous expansion. Gaseous work is already comprehended by ΔH.) Therefore, $w_{max} = \Delta G$, the free energy change for the process, and this equation is seen to be identical with eq. 6.3.

We are now able to calculate values of ΔG from simple cell measurements. It is not necessary to perform actual calorimetric experiments. It is often more accurate to calculate the electrical energy from equilibrium cell potentials. As in Section 6.3, the maximum electrical work from the cell is

$$w_{max} = -\text{charge} \times \text{reversible e.m.f.},$$

and so

$$\boxed{\Delta G = -nFE}. \qquad (9.1)$$

If all reactants and products are in their standard states,

$$\boxed{\Delta G^0 = -nFE^0}. \qquad (9.2)$$

Here, n is the number of moles of electrons, or equivalents involved in the reaction, and F, the Faraday constant, is a convenient unit of charge, equal to 96 487 coulombs/equiv.

9.3 Standard Electrode Potentials

There are many different half cells. From these one can construct many more cells, and to tabulate them all would be a huge task. However, we have seen that values of ΔU, ΔH, ΔS, and ΔG for a few simple reactions may be combined to give new values for other reactions. Similarly, data from a few cells may be combined together to yield data for many other possible cells. *The cells chosen for tabulation are those involving the standard hydrogen electrode* (SHE). What follows is based on the sign convention of the *International Union of Pure and Applied Chemistry* (IUPAC), agreed at the Stockholm Convention of 1953. (N.B. Other sign conventions have been used, and one should take care when consulting books and journals as to which one is being used. Many American books, in particular, use a convention in which opposite signs are used for electrode potentials. Always check the electrode potential of a common reactive metal such as sodium. If it is negative, then the convention is consistent with the IUPAC scheme.)

To write down the diagram $Pt,H_2 \mid H^+ \parallel Zn^{2+} \mid Zn$ necessarily implies a reaction with H_2 as reactant;

$$\tfrac{1}{2} H_2 + \tfrac{1}{2} Zn^{2+} \rightleftharpoons H^+ + \tfrac{1}{2} Zn. \tag{9.3}$$

In fact for standard states of unit activity, an e.m.f. of 0.763 V is observed, with the zinc negatively charged. This justifies the statement that *the standard electrode potential for the* $Zn^{2+} \mid Zn$ *half cell,* $E^0(Zn^{2+} \mid Zn) = -0.763$ V. The sign is that of the charge found on the right-hand side of the cell. Using eq. 9.2,

$$\Delta G^0 = -nFE^0$$
$$= -1(96\,487)(-0.763) = \mathbf{+73.63\ kJ/mol\ H^+\ formed.}$$

This result (ΔG^0 positive) confirms our belief that zinc displaces hydrogen ions from acid solution, rather than vice versa.

In the same manner, we interpret the fact that $E^0(Cl^- \mid AgCl, Ag) = +0.2225$ V to imply the following:

(a) $Pt, H_2 \mid H^+ \parallel Cl^- \mid AgCl, Ag^{\oplus}$

(b) $\tfrac{1}{2} H_2 + AgCl \rightleftharpoons H^+ + Cl^- + Ag$ (9.4)

(c) $\Delta G^0 = -1(96\,487)(+0.2225) = \mathbf{-21.47\ kJ/mol\ AgCl.}$

Notice that for any half cell which attains a *positive* charge when measured against a standard hydrogen electrode, the value of ΔG^0 will be *negative.* This shows that the associated reaction will be spontaneous. Reactions involve transfer of *one* electron for simplicity; and half cells are written as in the cell diagram.

Consider now a cell in which these two half cells are combined, such as $Zn \mid Zn^{2+} \parallel Cl^- \mid AgCl, Ag$. The cell reaction is written down, starting as before with the metal (or other reduced species) appearing on the left of the cell diagram (Zn in this case). (In cases where this is not straightforward, remember that product formation in the cell reaction must correspond to *movement in the cell of positive charge to the right.*) For this cell,

$$\tfrac{1}{2} Zn + AgCl \rightleftharpoons \tfrac{1}{2} Zn^{2+} + Cl^- + Ag.$$

Then E^0 (cell) is calculated by the 'right minus left' convention of eq. 1.1:

$$E^0\,(\text{cell}) = E^0\,(\text{right half cell}) - E^0\,(\text{left half cell})$$
$$= +0.2225 - (-0.763) = \mathbf{+0.9855\ V.}$$

The positive sign is the sign of the charge that turns up spontaneously on the right side of the cell. This means that the cell reaction does in fact proceed spontaneously as written. This is confirmed by the negative value of ΔG^0 which now emerges. Using eq. 9.2,

$$\Delta G^0 = -nFE^0 = -1(96\,487)(+0.9855)$$
$$= \mathbf{-95.09\ kJ/mol\ AgCl.}$$

These difficult-to-remember conventions of sign and so on may be represented visually; the double-headed arrows link cause with effect.

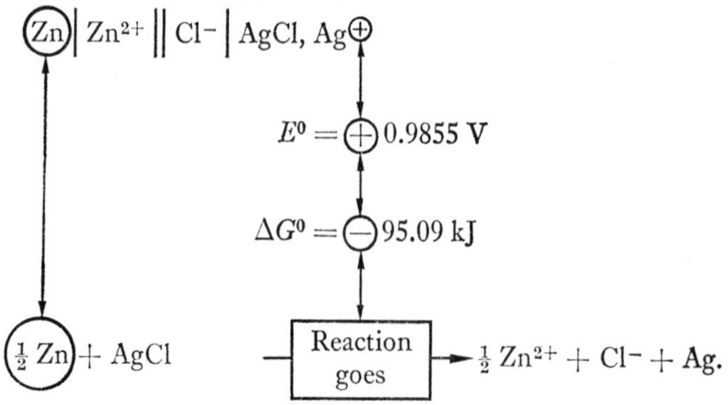

The four double-headed arrows have the following significance. (1) The reduced state on the left-hand side of the cell diagram begins the reaction equation. (2) The sign on the right of the cell diagram is the sign of the value of E^0, and (3)ΔG^0 is accordingly negative. (4) The reaction therefore proceeds from left to right. For other cells, positive and negative signs may of course be interchanged. Table 9.1 gives a selection of standard electrode potentials. An excellent collection of *oxidation potentials* (which are of the opposite sign) has been made by Latimer [3].

TABLE 9.1 *A selection of standard electrode potentials, E^0. Potentials apply to the cell* Pt, H_2 | H^+ || M^+ | M; *positive values imply spontaneity, with all materials at unit activity*

Electrode	Cell reaction	E^0
Li^+ | Li	$\frac{1}{2}H_2 + Li^+ \rightleftharpoons Li + H^+$	−3.025 volts
K^+ | K	$\frac{1}{2}H_2 + K^+ \rightleftharpoons K + H^+$	−2.925
Ca^{2+} | Ca	$\frac{1}{2}H_2 + \frac{1}{2}Ca^{2+} \rightleftharpoons \frac{1}{2}Ca + H^+$	−2.87
Na^+ | Na	$\frac{1}{2}H_2 + Na^+ \rightleftharpoons Na + H^+$	−2.714
Mg^{2+} | Mg	$\frac{1}{2}H_2 + \frac{1}{2}Mg^{2+} \rightleftharpoons \frac{1}{2}Mg + H^+$	−2.37
Zn^{2+} | Zn	$\frac{1}{2}H_2 + \frac{1}{2}Zn^{2+} \rightleftharpoons \frac{1}{2}Zn + H^+$	−0.763
Fe^{2+} | Fe	$\frac{1}{2}H_2 + \frac{1}{2}Fe^{2+} \rightleftharpoons \frac{1}{2}Fe + H^+$	−0.440
Cd^{2+} | Cd	$\frac{1}{2}H_2 + \frac{1}{2}Cd^{2+} \rightleftharpoons \frac{1}{2}Cd + H^+$	−0.403
Sn^{2+} | Sn	$\frac{1}{2}H_2 + \frac{1}{2}Sn^{2+} \rightleftharpoons \frac{1}{2}Sn + H^+$	−0.136
H^+ | H_2, Pt		0.00
Cl^- | AgCl, Ag	$\frac{1}{2}H_2 + AgCl \rightleftharpoons Ag + H^+Cl^-$	+0.2225
Cl^- | Hg_2Cl_2, Hg	$\frac{1}{2}H_2 + \frac{1}{2}Hg_2Cl_2 \rightleftharpoons Hg + H^+Cl^-$	+0.2675
Cu^{2+} | Cu	$\frac{1}{2}H_2 + \frac{1}{2}Cu^{2+} \rightleftharpoons \frac{1}{2}Cu + H^+$	+0.344
Fe^{3+}, Fe^{2+} | Pt	$\frac{1}{2}H_2 + Fe^{3+} \rightleftharpoons Fe^{2+} + H^+$	+0.771
Ag^+ | Ag	$\frac{1}{2}H_2 + Ag^+ \rightleftharpoons Ag + H^+$	+0.799
Br^- | Br_2, Pt	$\frac{1}{2}H_2 + \frac{1}{2}Br_2 \rightleftharpoons H^+ Br^-$	+1.066
Cl^- | Cl_2, Pt	$\frac{1}{2}H_2 + \frac{1}{2}Cl_2 \rightleftharpoons H^+Cl^-$	+1.359
Ce^{4+}, Ce^{3+} | Pt	$\frac{1}{2}H_2 + Ce^{4+} \rightleftharpoons Ce^{3+} + H^+$	+1.61
	IN GENERAL: $\frac{1}{2}H_2$ + (OXIDIZED STATE) $\rightleftharpoons$ (REDUCED STATE) + H^+	

Example 9.1 • Find the equilibrium constant at 25°C for the reaction

$$Hg + Cl^- + Ce^{4+} \longrightarrow \tfrac{1}{2} Hg_2Cl_2 + Ce^{3+}.$$

This reaction would occur in the cell Hg, Hg_2Cl_2 | Cl^- || Ce^{4+}, Ce^{3+} | Pt, and from Table 9.1,

$$E^0\,(\text{cell}) = E^0\,(\text{right}) - E^0\,(\text{left}) = +1.61 - 0.27 = \mathbf{+1.34\ V.}$$

Then $\Delta G^0 = -nFE^0 = -1(96\,487)(+1.34) = \mathbf{-129.3\ kJ/mol\ Hg.}$
Further, $\Delta G^0 \quad = -2.303\ RT \log K_a$, whence $K_a = \mathbf{10^{23}}$.

9.4 Variation of Cell e.m.f. with Activity

We are now in a position to calculate the standard e.m.f. of a cell, E^0, for standard state conditions; but unit activities are seldom encountered. There is a mass action effect, and we shall now investigate this.

The general reaction isotherm is

$$\Delta G = \Delta G^0 + RT \ln \frac{a_P^p\, a_Q^q \ldots}{a_A^a\, a_B^b \ldots}. \qquad (7.13)$$

If ΔG and ΔG^0 refer to a cell reaction, eqs. 9.1 and 9.2 may be enlisted and the activity quotient written Q, to give

$$\boxed{E = E^0 - \frac{RT}{nF} \ln Q}\,. \qquad (9.5)$$

This important expression is known as the Nernst equation. Now the reaction equation for cells which include the SHE is

$$\tfrac{1}{2} H_2 + \text{Oxidized States} \rightleftharpoons \text{Reduced States} + H^+,$$

so

$$E = E^0 - \frac{RT}{nF} \ln \frac{a_{H^+}\, a_{\text{Reduced States}}}{a_{H_2}^{\frac{1}{2}}\, a_{\text{Oxidized States}}}.$$

However, for cells which include the SHE, $a_{H^+} = a_{H_2} = 1$, and so in general, *for electrode potentials* (after inverting the ln term, and reversing the sign),

$$\boxed{E = E^0 + \frac{RT}{nF} \ln \frac{a_{\text{Oxidized States}}}{a_{\text{Reduced States}}}}\,. \qquad (9.6)$$

We must remember that the activities appear as they would in the activity quotient. As before, F is the Faraday constant, and n the number of moles of electrons involved in the process. If decadic logarithms are used, and the temperature is 298 K, a value of 0.059 15 may be used for $RT(2.303)/F$.

Example 9.2 The following cell is used to determine ion activities in chloride solutions

$$Pt \mid Fe^{2+}\ (a = 0.2),\ Fe^{3+}\ (a = 0.05) \parallel Na^+Cl^-\ (a = ?) \mid AgCl, Ag.$$

This would involve two vessels, linked by a salt bridge; in the left-hand one would be a ferrous-ferric salt solution, with a platinum electrode, and in the right, the sodium chloride solution with a silver-silver chloride electrode. The *mean ionic activities* (discussed later) are indicated.

E for a particular chloride solution is -0.430 V. What is its activity?

The cell reaction is

$$Fe^{2+} + AgCl \rightleftharpoons Fe^{3+} + Cl^- + Ag.$$

(Note that the equation begins with the reduced state (Fe^{2+}) in place of a metal.) Now E^0 (cell) $= E^0$ (right) $- E^0$ (left) $= 0.2225 - (+0.771) = \mathbf{-0.5485\ V}$. Using eq. 9.5, we find that

$$-0.430 = -0.5485 - .059\,15 \log \frac{a_{Fe^{3+}}\, a_{Cl^-}\, a_{Ag}}{a_{Fe^{2+}}\, a_{AgCl}}. \tag{9.7}$$

The activities of silver and silver chloride, being crystalline solids, are each unity. Further arithmetic shows that

$$\log \frac{a_{Cl^-}(0.05)}{0.2} = -2.00$$

whence

$$a_{Cl^-} = \mathbf{0.04.}$$

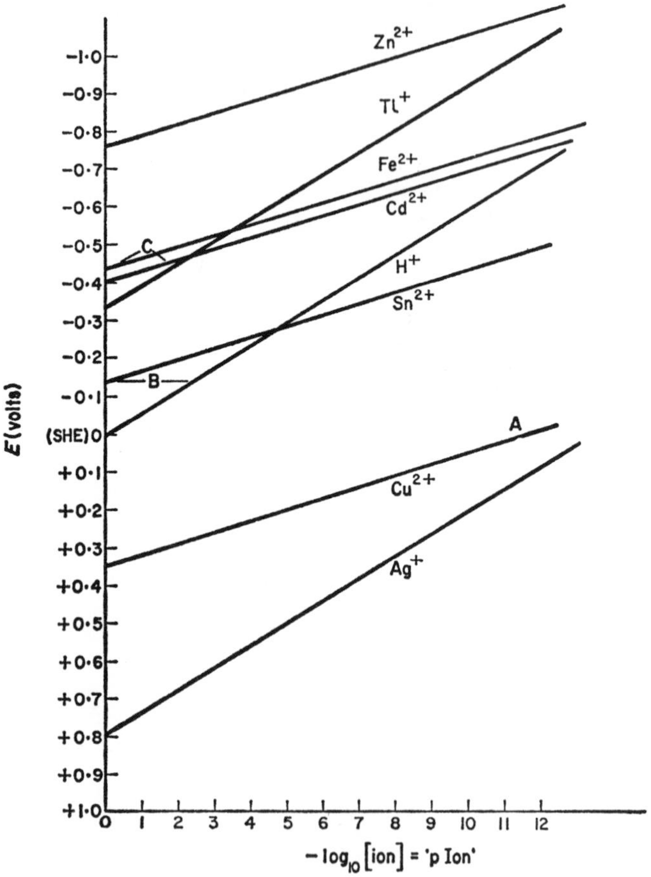

FIG. 9.4 The electrode potentials of some common metals, compared with the SHE, at various concentrations. Activity and concentration are assumed equal throughout, and $T = 298$ K.

The way in which cell e.m.fs vary with concentration is shown clearly by means of a graph. Figure 9.4 shows the e.m.fs of a number of metal electrodes against the SHE, that is, of cells of the type Pt, H_2 (1 atm) | H^+ ($a = 1$) || M^{n+} | M. Notice that for monovalent ions, the e.m.f. varies by 0.059 volts every *p Ion* unit (every factor of ten in concentration). For divalent ions, the difference is $0.059/n = 0.0295$ V. It is interesting to observe that:

(a) Copper is stable with respect to the SHE *unless* the cupric ion concentration $[Cu^{2+}]$ drops below 10^{-12}, *when it will tend to dissolve* (Point A).
(b) Tin will dissolve in acid to give a molar Sn^{2+} solution *if the acid is 10^{-2} molar or stronger*, but not otherwise (Point B).
(c) Ferrous ions will displace Cd^{2+}, *or vice versa*, depending on their relative concentrations (Point C).

Such charts are of great use for electrolytic refining, plating, and corrosion studies.

9.4.1 Ionic Activities

Complete cell equations never contain one ion at a time, so individual ionic activities cannot be determined. For example, eq. 9.7 shows that $a_{Fe^{3+}}$ occurs together with a_{Cl^-} and $a_{Fe^{2+}}$, and this situation is typical of others.

Let us focus attention on the simple 1 : 1 electrolyte H^+Cl^-. Its activity a, *as a compound*, may be found from colligative properties, but the activities of the ions, a_+ and a_-, are not determinable separately. These ionic activities must first be carefully defined in terms of a reference state with unit fugacity; as before, $a_+ = f_+/f_+^0$. The reference states are defined *such that*, for the equation

$$\begin{array}{ccc} HCl \rightleftharpoons & H^+ + & Cl^- \\ (a) & (a_+) & (a_-) \end{array}$$

the free energy change is zero, and the equilibrium constant $K = 1$. Therefore $a_+a_-/a = 1$ and $a = a_+a_-$. Because a_+ and a_- cannot be separately determined, we define the **mean ionic activity** as $a_\pm = (a_+.a_-)^{1/2}$. Also,

$$\gamma_\pm = (\gamma_+\gamma_-)^{1/2}$$

and

$$m_\pm = (m_+m_-)^{1/2}. \tag{9.8}$$

We now have activity, molality, and activity coefficient, defined in terms of the geometric mean of the ionic values, with, as usual,

$$a_\pm = \gamma_\pm m_\pm.$$

(This applies to 1 : 1 electrolytes only, but similar arguments apply to polyfunctional ions.) Note that if the electrolyte is substantially dissociated, then $m_+ = m_- = m$, and $m_\pm = m$.

9.4.2 Analysis of e.m.f. Data to Find E^0

Standard electrode potentials are based on electrolytes of unit activity, with a 'hypothetical ideal molality of one'. Raw laboratory data arrive as simple e.m.f. values versus molality. How do we find E^0?

Some data of this kind for the cell

$$Pt, H_2 \text{ (1 atm)} \mid H^+Cl^- (m) \mid Hg_2Cl_2, Hg$$

are given in Table 9.2. For this cell, the reaction and eq. 9.5 inform us that

$$\begin{array}{lccccc} & \tfrac{1}{2} H_2 (g) + & \tfrac{1}{2} Hg_2Cl_2 (c) \longrightarrow & H^+ + & Cl^- + & Hg (l), \\ \text{activity} = & 1 & 1 & a_+ & a_- & 1, \end{array}$$

and

$$E = E^0 - \frac{RT}{F} \ln a_+a_-.$$

TABLE 9.2 *E.M.F. data at various molalities for the cell composed of the hydrogen and calomel half cells* (25 °C)

Molality, m, mol/kg water	E, volt	$m^{1/2}$	$E + 0.118\,3 \log m$
0.07508	0.4119	0.2740	0.2789
0.03769	0.4452	0.1941	0.2768
0.01887	0.4787	0.1374	0.2747
0.00504	0.5437	0.0710	0.2719
0.00200	0.5900	0.0447	0.2707

We have already decided that $a_+a_- = a_\pm^2$, and $a_\pm = \gamma_\pm m_\pm = \gamma_\pm m$,

so
$$E = E^0 - \frac{2RT}{F} \ln \gamma_\pm m$$
$$= E^0 - 2(.059\,15) \log \gamma_\pm m.$$

Thus
$$E^0 = \underbrace{E + 0.118\,3 \log m}_{\text{Approximately constant}} + \underbrace{0.118\,3 \log \gamma_\pm}_{\text{Correction term, approaching zero at low concentration}}. \quad (9.9)$$

It seems that E^0 is approximately equal to the expression $(E + 0.118\,3 \log m)$, and that it approaches it more and more closely as m tends to zero, and $\gamma_\pm$ tends to one. This term is presented in column four of Table 9.2. In the approach to electrolytic solutions pioneered by Debye and Hückel, it was shown that $\log \gamma_\pm$ varied linearly with $m^{1/2}$, for weak solutions. This means that $(E + 0.118\,3 \log m)$ should show a similar variation; this suspicion is confirmed by Fig. 9.5, which

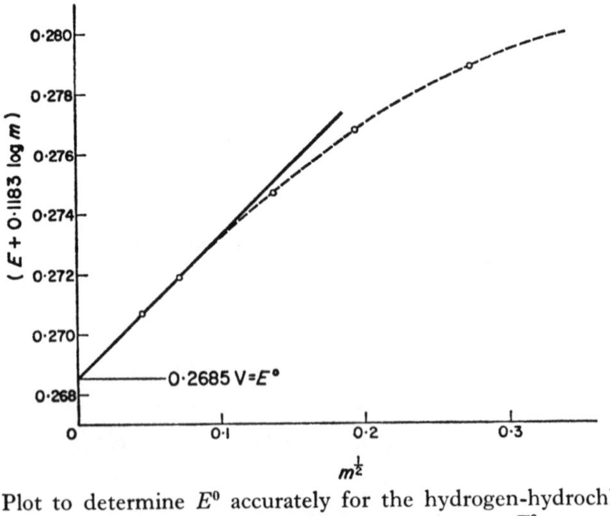

FIG. 9.5 Plot to determine E^0 accurately for the hydrogen-hydrochloric acid-calomel cell. The term $(E + 0.118\,3 \log m)$ approximates E^0 more closely as concentrations approach zero.

shows that $(E + 0.118\,3 \log m)$ varies approximately linearly with $m^{1/2}$. The resulting value of E^0, obtained by extrapolation to ideal conditions, is 0.268 5 V, and is in good agreement with other determinations. Once E^0 is known, it is a simple matter to calculate the activity coefficient, $\gamma_{\pm}$, at a variety of concentrations, from eq. 9.9. At a molality of 0.075 08, $\gamma_{\pm}$ is 0.823. This should be checked and other data calculated.

9.5 Variation of e.m.f. with Temperature

One of the most rewarding aspects of electrochemistry is that ΔS values spring directly from studies of the temperature variation of e.m.f.

Equation 10.1, to be derived in Chapter 10, describes the temperature variation of ΔG:

$$\left(\frac{\mathrm{d}(\Delta G)}{\mathrm{d}T}\right)_P = -\Delta S. \qquad (10.1)$$

If we substitute for ΔG from eq. 9.2 which is $\Delta G = -nFE$, we have

$$\left(\frac{\mathrm{d}(-nFE)}{\mathrm{d}T}\right)_P = -\Delta S$$

whence

$$\boxed{nF\left(\frac{\mathrm{d}E}{\mathrm{d}T}\right)_P = \Delta S}\,, \qquad (9.10)$$

and also

$$\boxed{nF\left(\frac{\mathrm{d}E^0}{\mathrm{d}T}\right)_P = \Delta S^0}\,. \qquad (9.11)$$

This provides a basically simple means of ΔS determination.

Example 9.3 King and Cobble [4] find that for the cell in which the reaction

$$\tfrac{1}{2}\,H_2\,(g) + HReO_4\,(aq) \longrightarrow ReO_3\,(c) + H_2O\,(l)$$

takes place, $\mathrm{d}E^0/\mathrm{d}T = -(1.21)10^{-3}$ V/K.

If S^0_{298} ($HReO_4$ (aq)) = 202.1 J/mol K, what is the molar entropy of ReO_3 (c)?

Using eq. 9.10, $\Delta S^0 = -1(96\,487)(1.21)10^{-3}$

$= \mathbf{-116.7\ J/mol\ K.}$

Using tabulated values of S^0 for H_2 and H_2O, we find that

$$S^0_{298}\,(ReO_3\,(c)) = -116.7 - 69.9 + 65.3 + 202.1$$

$$= \mathbf{+80.8\ J/mol\ K.}$$

Example 9.4 Worrell [5] at the University of California has studied cells with solid electrolytes which are O^{2-} ionic conductors. Niobium oxides were studied by use of the cell

$$^{\ominus}Nb,\, NbO \mid ThO_2\,(Y_2O_3\ 2\%) \mid NbO,\, NbO_2{}^{\oplus}.$$

The electrode reactions are

$$Nb + O^{2-} \longrightarrow NbO + 2e \quad \text{and} \quad NbO_2 + 2e \longrightarrow NbO + O^{2-}$$

which give overall, $Nb + NbO_2 \rightleftharpoons 2NbO$.

Values of e.m.f. at various temperatures are given in Table 9.3.

TABLE 9.3 *Values of e.m.f. from* 1050–1300 K *for a solid electrolyte cell*

E, mV	T K	E, mV	T K
213	1054	202	1174
212	1086	200	1214
209	1106	198	1244
201.5	1160	194	1283

Figure 9.6 shows a graph of E versus T, from which a value for $\mathrm{d}E/\mathrm{d}T$ of $(-9.0)10^{-5}$ volt/K is obtained. Using eq. 9.10, this gives

$$\Delta S = 2(96\,487)(-9.0)10^{-5} = \mathbf{-17.4\ J/mol\ K.}$$

In addition, values of ΔG are available at any temperature, calculated as $-nFE$. At 1200 K,

$$\Delta G = -2(96\,487)(0.201) = \mathbf{-38.8\ kJ/mol.}$$

Then

$$\Delta H = \Delta G + T\Delta S = -38\,800 + 1200(-17.4) = \mathbf{-59.70\ kJ/mol.}$$

Finally all the available thermodynamic information may be summarized in terms of the equation

$$\Delta G = \Delta H - T\Delta S$$
$$= \mathbf{-59\,700 + 17.4\,T\,(joule/mol).}$$

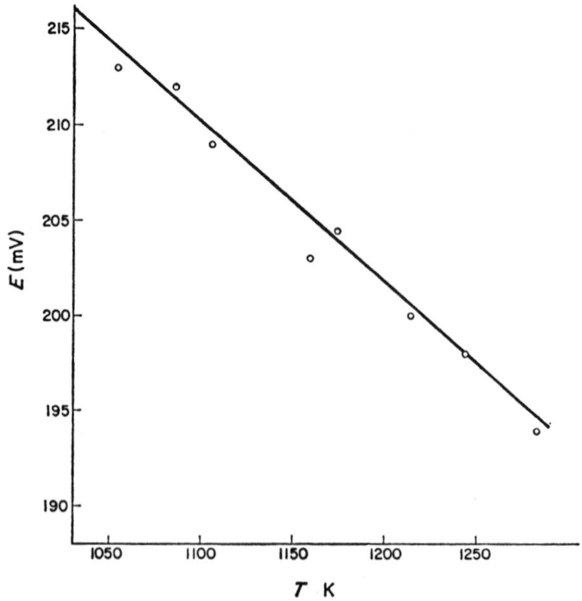

FIG. 9.6 Graph of e.m.f. against temperature for the 'Worrell' cell, from 1054 to 1283 K. The cell is solid throughout.

REFERENCES

[1] MOORE, W. J. *Physical Chemistry*, 4th edition, Longmans, London (1963).
[2] POTTER, E. C. *Electrochemistry*, Cleaver-Hume, London (1961).
[3] LATIMER, W. *Oxidation Potentials*, Prentice-Hall, Englewood Cliffs, N.J. (1952).
[4] KING, J. P. and COBBLE, J. W. *J. Am. chem. Soc.* **79,** 1559 (1957).
[5] WORRELL, W. L. *Symposium on Thermodynamics*, 131, International Atomic Energy Agency, Vienna (1966).

PROBLEMS

9.1 An iron penknife blade may be 'copper plated' in $CuSO_4$ solution, by displacement of Cu^{2+} ions with Fe^{2+} ions. Assuming that enough penknife blades are available, how low can the Cu^{2+} activity be taken at 25°C if the initial activity is one?

9.2 The compound Ni_2O_3 is a strong oxidizing agent, and is used in the Edison storage cell, which is represented as

$$Fe, FeO \mid KOH\ (aq) \mid Ni_2O_3, NiO^{\oplus}.$$

The cell reaction is

$$Fe + Ni_2O_3 \longrightarrow FeO + 2\ NiO.$$

(a) Write down the half-cell reactions.
(b) If the cell e.m.f. is +1.27 volts, what is the free energy of formation of Ni_2O_3? (Data available from Table 2.1.)
(c) What effect does KOH activity have on the equilibrium e.m.f.? on the current?

9.3 The following reactions have been used for fuel cells. Write down a cell diagram and electrode reactions for each.

$$2\ CuBr_2 \rightleftharpoons 2\ CuBr + Br_2$$
$$PbI_2 \rightleftharpoons Pb + I_2$$
$$2\ LiH \rightleftharpoons 2\ Li + H_2 \qquad \text{(care!)}$$

If $\Delta G^0_f\,(LiH) = -69.96$ kJ/mol, what equilibrium potential is expected for the third cell?

9.4 Silver ions oxidize ferrous ions according to the equation:

$$Ag^+\,(aq) + Fe^{2+}\,(aq) \longrightarrow Ag\,(c) + Fe^{3+}\,(aq).$$

The equilibrium constant at 298 K is $K_a = 7.810$. This reaction may also be studied as a cell, such as

$$Pt \mid Fe^{2+}, Fe^{3+} \mid NH_4NO_3 \text{ bridge} \mid Ag^+\ NO_3^- \mid Ag.$$

Determine
(a) the half-cell reactions,
(b) the polarity, and
(c) the expected e.m.f. of the cell with unit activity electrolytes.

9.5 The following cell is studied experimentally:

$$Pt, H_2\,(1\text{ atm}) \mid H^+Cl^-\,(a = 1) \mid H_2\,(0.2\text{ atm}), Pt.$$

What is the cell reaction? What polarity do you suggest? Assuming that $a_{H_2} = P_{H_2}$, calculate the cell e.m.f.

9.6 The *glass electrode* is sensitive to the activity of hydrogen ions. If used in conjunction with a calomel reference electrode and a KCl salt bridge, the equations

$$E = E^0 + \frac{RT}{F} \ln a_{H^+},$$

and

$$E = E^0 - \frac{2.303RT}{F} \text{pH} \quad \text{apply,}$$

where pH is defined here as being equal to $-\log a_{H^+}$. The following data obtained at 25°C from such a cell enable one to derive a value for F, the Faraday constant, graphically. Express your answer in coulomb/equiv.

pH	4.01	6.49	6.99	9.15
E, V	−0.1245	−0.271	−0.3015	−0.430

9.7 Values of e.m.f. for the cell

$$\text{Pt, } H_2 \text{ (1 atm)} \mid H^+Br^- \text{ (m)} \mid \text{AgBr, Ag}$$

were determined at various molalities to be as follows, at 25°C:

m(mol/kg H_2O)	E(V)
$(1.26)10^{-4}$	0.5330
$(1.77)10^{-4}$	0.5162
$(4.17)10^{-4}$	0.4721
$(1.10)10^{-3}$	0.4228

Determine E^0 (Br^- | AgBr, Ag) accurately by a graphical method.

9.8 Write the electrode reactions, and the total cell reaction for

$$\text{Pt, } H_2 \text{ (1 atm)} \mid \text{HCl in MeOH} \mid \text{AgCl, Ag.}$$

At 25°C, data for this cell are

c (mol/l)	E (Volt)
0.002 110	0.304 2
0.004 265	0.271 3
0.007 749	0.244 6
0.013 53	0.220 8

From an appropriate plot of these data calculate E^0, the standard electrode potential of the silver-silver chloride electrode in methanol solvent.

9.9 A student study of the cell Pt, H_2 (1 atm) | H^+Cl^- (c = 1) | AgCl, $Ag^{\oplus}$ produced the following data of e.m.f. versus temperature:

T°C	e.m.f. (mV)	T°C	e.m.f. (mV)
26	206	40	197
28	205	42	196
30	204	44	194
32	202	46	193
34	201	48	191
36	199.5	50	190
38	198	52	188.5

Determine ΔS for the cell reaction, by graphical means and compare with ΔS^0_{298} calculated from tables.

9.10 The cell represented by Pb, PbI_2 | K^+I^- solution | AgI, Ag has an e.m.f. of +0.207 8 V at 25°C, and $(dE/dT)_P = -(1.88)10^{-4}$ V/K. Write down the cell reaction, and calculate ΔG, ΔS, and ΔH. It will be found that all components appearing in K_a are solids; why therefore is not $K = 1$, $\Delta G = 0$?

9.11 At 25°C the e.m.f. of the cell

$$\text{Pb} \mid \text{Pb}^{2+} \mid 5 \text{ per cent Pb amalgam} \mid \text{Hg}$$

is $(5.7)10^{-3}$ V and its temperature coefficient is $(1.6)10^{-4}$ V/K.

Calculate: (a) the entropy change when current is drawn reversibly from the cell, and
(b) the associated enthalpy change.

9.12 The Weston standard cell provides 1.0186 V at 20°C, and

$$(dE/dT)_P = -(4.06)10^{-5} \text{ V/K}.$$

Calculate ΔG_{293} for the process:

$$\tfrac{1}{2}\,\text{Cd} + \tfrac{1}{2}\,\text{Hg}_2\text{SO}_4 \longrightarrow \tfrac{1}{2}\,\text{CdSO}_4\,(\text{c}) + \text{Hg}$$

which is the spontaneously occurring change. Find also ΔS; why is the Weston cell so well suited to be a standard?

10

Free Energy and Industrial Processes

At the start of this chapter, we shall attempt to describe accurately the changes of ΔG with temperature. This will require careful computation of ΔC_P. Then, using free energy data, we shall discuss the outcome of three chemical processes, all operated on an industrial scale. These are the direct synthesis of ethanol from ethylene, which has now largely superseded the fermentation processes, the Pidgeon process for the production of magnesium, and the process used for titanium manufacture. Ellingham diagrams will be introduced and applied, and a practical assessment made of the importance of activity coefficients.

10.1 Free Energies as a Function of Temperature

We have often used the equation defining free energy changes, as

$$\Delta G = \Delta H - T\Delta S. \tag{6.3}$$

As a rule, when calculating ΔG at temperatures other than 298 K, we have made the assumption that ΔH and ΔS are independent of temperature. That is, we have calculated free energy changes at any temperature T as

$$\Delta G^0_T = \Delta H^0_{298} - T\Delta S^0_{298},$$

and although this is adequate for small temperature ranges, significant errors are introduced when, for example, high temperature metallurgical processes are described. Although individual variations in ΔH and ΔS have been accounted for in theory (e.g., Section 8.1) no general expression for ΔG at any temperature has been attempted. This we shall now remedy.

10.1.1 The Gibbs-Helmholtz Equation

We start with eq. 7.1, which asserts that

$$\mathrm{d}G = V\,\mathrm{d}P - S\,\mathrm{d}T. \tag{7.1}$$

In constant pressure situations, $\mathrm{d}P = 0$, and we may write

$$\left(\frac{\mathrm{d}G}{\mathrm{d}T}\right)_P = -S.$$

Alternatively, for a reaction, we could modify this to read

$$\boxed{\left(\frac{\mathrm{d}(\Delta G)}{\mathrm{d}T}\right)_P = -\Delta S} \tag{10.1}$$

This expression has very great usefulness, and has already been exploited in the previous chapter. On introducing this into eq. 6.3, we have one form of the *Gibbs-Helmholtz equation,*

$$\boxed{\Delta G = \Delta H + T\left(\frac{\mathrm{d}(\Delta G)}{\mathrm{d}T}\right)_P}. \tag{10.2}$$

In the discussion which follows, we shall find an alternative expression preferable.

Let us now differentiate $\Delta G/T$, with respect to temperature. From the usual rule for differentiating a quotient, we find that

$$\frac{\mathrm{d}(\Delta G/T)}{\mathrm{d}T} = \frac{T\frac{\mathrm{d}(\Delta G)}{\mathrm{d}T} - \Delta G}{T^2}.$$

If we now substitute for ΔG from eq. 10.2, we have

$$\frac{\mathrm{d}(\Delta G/T)}{\mathrm{d}T} = -\frac{\Delta H}{T^2}. \tag{10.3}$$

This is merely an alternative expression for eq. 10.2. However, as they stand, neither of these alternatives can help us to find high temperature values of ΔG, accurately. Clearly, we must integrate one of these expressions.

10.1.2 The Integrated Form of the Gibbs-Helmholtz Equation

To integrate eq. 10.3, we start by writing it as

$$\int \mathrm{d}(\Delta G/T) = \int(-\Delta H/T^2)\,\mathrm{d}T. \tag{10.4}$$

We are now obliged to express ΔH as a function of temperature. From Section 3.2 we recall that $(\mathrm{d}H/\mathrm{d}T)_P = C_P$, and that heat capacities were expressed as simple polynomials in T, of the form $C_P = a + bT + cT^2 + dT^3$. For reactions, we write

$$\begin{aligned}\left(\frac{\mathrm{d}(\Delta H)}{\mathrm{d}T}\right)_P &= \Delta C_P \\ &= \Delta a + \Delta bT + \Delta cT^2 + \Delta dT^3\end{aligned} \tag{10.5}$$

where Δ has the usual meaning of 'products minus reactants'. On integrating,

$$\int \mathrm{d}(\Delta H) = \int(\Delta a + \Delta bT + \Delta cT^2 + \Delta dT^3)\,\mathrm{d}T$$

$$\Delta H_2 - \Delta H_1 = \left[\Delta aT + \Delta bT^2/2 + \Delta cT^3/3 + \Delta dT^4/4\right]_{T_1}^{T_2}.$$

In general,

$$\boxed{\Delta H_T = \Delta H_0 + \Delta aT + \Delta bT^2/2 + \Delta cT^3/3 + \Delta dT^4/4} \tag{10.6}$$

where ΔH_0 is the value of ΔH_T when $T = 0$ K. One must be careful to notice the difference between the zero as superscript, meaning standard state, and as subscript, meaning zero temperature. This expression for ΔH_T may now be inserted into eq. 10.4, to give, on integration

$$\boxed{\Delta G_T = \Delta H_0 - \Delta aT\ln T - \Delta bT^2/2 - \Delta cT^3/6 - \Delta dT^4/12 + IT}. \tag{10.7}$$

The term I is an integration constant, yet to be determined. It should be stressed that this impressive equation can be taken as seriously or as lightly as the circumstances demand. In some cases, the accuracy will only justify an averaged value of Δa, later terms being neglected. However, even when simplified to this extent, and especially when taken rigorously, it can involve a great deal of work. Suppose we know ΔG_{298}, ΔH_{298}, and the expressions for C_P for each reactant and product. We need to know ΔH_0, and I before proceeding with the ΔG_T evaluation. The scheme of calculation would be somewhat as follows:

(a) determine Δa, Δb, and so on from the C_P expressions;
(b) use eq. 10.6, the known value for ΔH_{298}, and $T = 298$, and calculate ΔH_0;
(c) use eq. 10.7, the value of ΔG_{298}, and ΔH_0, with $T = 298$, and find I;
(d) calculate ΔG at any desired temperature.

10.1.3 Tabulated Forms of Free Energy

Surely, the persistent optimist will say, 'there must be an easier way'. There are two developments which we might examine, before moving to some practical applications. The first possibility is to tabulate ΔG^0 over a range of temperatures, say every 200°, for all formation reactions, and then to find the desired ΔG^0 values by interpolation. This is a good idea in theory, but in practice is simply not accurate enough for detailed calculations of equilibrium constants. However, it is a very useful aid in the preliminary assessment of untried processes. We shall say more about this kind of approach later, when discussing Ellingham diagrams. The second possibility is to find a function which involves ΔG^0, but does not change with temperature as violently as does ΔG^0 itself. It has been found that the *free energy function* (f.e.f.), $-(G^0 - H_0^0)/T$, which is assignable to individual compounds and elements rather than reactions, does just this. It has the additional advantage that it may be calculated from spectroscopic data, which are inherently reliable and accurate. The reader is referred to Lewis and Randall's book for tabulations of f.e.fs and further discussion of them [1].

10.2 The Synthesis of Ethanol

Ethanol has traditionally been made on a large scale from the fermentation of molasses and grains. The impressive rise of the petrochemicals industry, however, has considerably altered the pattern of the industry on both sides of the Atlantic. In 1950, only 6 per cent of the total UK ethanol output came from the ethylene hydration process. In 1960, the share had grown to nearly 70 per cent, and it is still rising. The process is based on the direct hydration of ethylene,

$$C_2H_4 + H_2O \rightarrow C_2H_5OH$$

at 1000 pounds pressure, and nearly 600 K, over a phosphoric acid catalyst. We shall discuss certain aspects of this reaction to exemplify detailed ΔG calculations, and also the application of activity coefficients (Ch. 7) to high pressure processes.

10.2.1 Equilibrium Calculations

We shall now follow the recipe given at the end of section 10.1.2. First, we find that the C_P functions are (data from Table 3.5)

$$C_P(C_2H_5OH) = 19.9 + (20.95)10^{-2}T - (10.37)10^{-5}T^2 + (20.04)10^{-9}T^3$$
$$C_P(C_2H_4) = 3.95 + (15.63)10^{-2}T - (8.339)10^{-5}T^2 + (17.66)10^{-9}T^3$$
$$C_P(H_2O) = 32.22 + (0.192)10^{-2}T + (1.054)10^{-5}T^2 - (3.594)10^{-9}T^3$$

and so $\Delta C_P = C_P(C_2H_5OH) - C_P(C_2H_4) - C_P(H_2O)$

$$= \mathbf{-16.29 + (5.125)10^{-2}T - (3.088)10^{-5}T^2 + (5.979)10^{-9}T^3.}$$

That is, in Eq. 10.6, $\Delta a = -16.29$, and so on.

Secondly, we need to know ΔH^0_{298} from enthalpy of formation data. This is found to be -45.756 kJ/mol, and so we can now use eq. 10.6, with $T = 298$ K. Thus

$$\Delta H^0_0 = \Delta H^0_{298} - \Delta a(298) - \Delta b(298^2)/2 - \Delta c(298^3)/3 - \Delta d(298^4)/4$$
$$= \mathbf{-42.919\ kJ/mol.}$$

Thirdly, we can use eq. 10.7 to find I, if we have ΔH^0_0 and ΔG^0_{298}. From the free energies of formation given in Table 2.1, $\Delta G^0_{298} = -8.125$ kJ/mol. As a result,

$$I = \Delta G^0_{298}/298 - \Delta H^0_0/298 + \Delta a \ln 298 + \Delta b(298)/2 + \Delta c(298^2)/6 + \Delta d(298^3)/12$$
$$= \mathbf{+31.062\ joule/mol\ K.}$$

Finally, our ΔG^0_T equation 10.7 now reads

$$\Delta G^0_T = -42\,919 - \Delta aT \ln T - \Delta bT^2/2 - \Delta cT^3/6 - \Delta dT^4/12 + 31.062T.$$

Taking all the plant parameters into account, the Shell Development Company [2] recommends operation at 572 K. In this case, we find

$$\Delta G^0_{572} = +26.543 \text{ kJ/mol},$$

and

$$K_a = (3.77)10^{-3}.$$

We must temper our initial disappointment at the small equilibrium constant by recalling that industrial processes often involve recycling of unreacted starting material. This depends on effective removal of what product there is, before the unreacted ethylene is put through the catalyst bed once again. In this case, washing of product mixture with liquid water is all that is required to remove the ethanol, and so commercial operation is practicable after all.

Before making accurate calculations of the equilibrium mixture, it is interesting to compare the value for ΔG^0_{572} obtained above with the one calculated from the simple $\Delta G^0 = \Delta H^0 - T\Delta S^0$ equation, and to compare them both with the definitive value obtained from experimental equilibrium measurements.

First, using

$$\Delta G^0_T = \Delta H^0_{298} - T\Delta S^0_{298},$$

we find that

$$\Delta G^0_{572} = +26.43 \text{ kJ/mol},$$

and

$$K_a = (3.86)10^{-3}.$$

The experimental value of K_a is [3] $(3.14)10^{-3}$, giving $\Delta G^0_{572} = +27.417$ kJ/mol. We shall adopt this value for future discussion. It will be noticed that although both calculated values of K_a are slightly higher than the experimental value, the more carefully calculated figure is at least the nearer of the two. All three values are close together, and we must conclude that ΔC_P is sufficiently small to be unimportant *in this case*. However, only by doing the calculation can we be sure of our derived information.

10.2.2 Use of Activity Coefficients

In Section 7.7, we discussed the derivation and use of the equation $\Delta G^0 = -RT \ln K_a$. In general, for each gas in a gas phase reaction, $f = \gamma p$. In the case where pressure is low, the fugacity approaches the pressure, and $\gamma = 1$. However, in the ethanol process, pressures are high, and we must express K_a as

$$K_a = \frac{f_{\text{alc}}}{f_{\text{e}} \cdot f_{\text{w}}} = \frac{p_{\text{alc}}}{p_{\text{e}} \cdot p_{\text{w}}} \cdot \frac{\gamma_{\text{alc}}}{\gamma_{\text{e}} \cdot \gamma_{\text{w}}}. \tag{10.8}$$

The subscripts alc, e, and w refer to alcohol, ethylene, and water, respectively. The term in pressures is usually written K_p, and the partial pressures are derived from the total pressure multiplied by mole fraction. The expression in activity coefficients, because of its similarity in form to K_p, is often abbreviated K_γ; it is *not* of course an equilibrium constant. In short, we may write $K_a = K_p K_\gamma$. The activity coefficients are determined from the generalized chart shown as Fig. 7.1, from known or estimated pressures. However, we know *only* K_a *and the total pressure* to start with, and must estimate starting values of each pressure in order to find values for γ in each case. We can then estimate K_p as K_a/K_γ, and calculate equilibrium pressures once more. If the new values for γ are appreciably different, then another iteration (calculation) is called for. The approach is similar to that of a golfer approaching his hole; he arrives by successive approximation. Let us see how this works out in practice.

The following parameters are quoted for the ethanol process:

Mole ratio of feedstock: Ethylene 10, water vapour 6
Temperature : 572 K
Total pressure : 66.3 atm.

The mole fraction and partial pressure situation is summarized as follows:

	C_2H_4	+ H_2O	$\rightleftharpoons C_2H_5OH$
Initial moles :	10	6	0
Initial partial pressure :	(66.3)10/16	(66.3)6/16	0
Moles at equilibrium :	$10 - x$	$6 - x$	x
Pressures at equilibrium :	$\dfrac{(10-x)66.3}{(16-x)}$	$\dfrac{(6-x)66.3}{(16-x)}$	$\dfrac{66.3x}{(16-x)}$

Therefore $$K_p = \frac{x(16-x)}{66.3(10-x)(6-x)},$$

and conversion of ethylene will be $100(x/10)$ per cent (for every 10 moles entering the plant, x react).

We begin by assuming $\gamma = 1$ for each gas, and $K_\gamma = 1$. Then $K_p = (3.14)10^{-3}$, and we can solve for x. This turns out to be 0.675, and so ethylene conversion is 100(0.675/10) or 6.75 per cent. Then $p_{\text{alc}} = 2.92$ atm, $p_{\text{e}} = 40.34$ atm, and $p_{\text{w}} = 23.04$ atm. This now enables us to look up γ values, using the data on critical points, and reduced temperatures and pressures shown in Table 10.1; this gives a value of $K_\gamma = 1.067$. On the second time around, using our improved value of $K_a/1.067 = (2.94)10^{-3}$ for K_p, we find new pressures and gammas, which give $K_\gamma = 1.070$. A third and final calculation gives $x = 0.636$, and the following results:

Alcohol	$\gamma = 0.99$	$p = 2.75$ atm	$f = 2.72$ atm
Ethylene	$= 0.99$	$= 40.41$ atm	$= 40.00$ atm
Water	$= 0.935$	$= 23.15$ atm	$= 21.64$ atm

$K_\gamma = \mathbf{1{\cdot}070}$, $K_p = \mathbf{(2{\cdot}94)}10^{-3}$ $K_a = \mathbf{(3{\cdot}14)}10^{-3}$.

TABLE 10.1 *Critical constants of ethanol, ethylene, and water*

	T_c K	P_c atm	T_r	P_r
Ethanol	516.3	63.1	1.108	0.048
Ethylene	282.9	50.9	2.022	0.792
Water	647.0	217.7	0.884	0.105

Because the gases do not behave ideally, the ethylene conversion is reduced from 6.75 per cent to 6.36 per cent, which represents a significant lowering in the expected yield.

It is a small change, however, when compared with other reactions; moreover, non-ideality may lead to an improvement in K_p, as in the ammonia synthesis. We find that K_p for the reaction

$$\tfrac{1}{2}\,N_2 + \tfrac{3}{2}\,H_2 \rightleftharpoons NH_3$$

is 0.006 6 at 450°C and 1 atm, but has risen to 0.013 for a total pressure of 600 atm. In this case, therefore, the yield will increase on two counts. Firstly, high pressure helps conversion to low volume products, and secondly, the gas imperfections happen to make a further change for the better.

It is interesting that the ethanol plant runs with about 5 per cent yield, which means that the catalyst (phosphoric acid on clay pellets) is not fully effective. The remaining 95 per cent of unreacted ethylene is recycled with an additional 5 per cent of fresh feed. On average, therefore, an ethylene molecule will make nineteen trips before reacting successfully on the twentieth.

10.3 Ellingham Diagrams

Calculations of ΔG_T^0 using eq. 10.3 are excessively time-consuming, but for high temperature processes such as occur in metal extraction and nuclear power plants data must be available over wide ranges of temperature. It is therefore necessary to take rigorous account of the temperature variation of ΔG^0, even though the variation often turns out to be approximately linear. It was H. J. T. Ellingham [4] who suggested that ΔG^0 data for many reactions of one class, say oxidations of metals, should be plotted against T in such a form that it could be seen at a glance if a certain metal would reduce the oxide of another. As Ellingham put it, '*It occurred to the present author that a useful purpose would be served by representing graphically the variation with temperature of the standard free energy of formation of compounds of a particular class (say the oxides) on a single diagram.*' It would provide, he said, '*what might be described as a ground plan of metallurgical possibilities.*' The discussion of such diagrams which follows owes much to the excellent monograph of Ives [5], to which further reference is recommended.

An example of such a chart is shown in Fig. 10.1 for oxide formation. Familiarity will come with practice with many different reactions, but for the present consider the line due to Zn/ZnO. Values of ΔG^0 for all the reactions shown on the diagram involve *one mole of oxygen*, in the formation of an oxide,

$$\frac{2x}{y}M + O_2 \rightarrow \frac{2}{y}M_xO_y;$$

here

$$2\,Zn + O_2 \rightarrow 2\,ZnO.$$

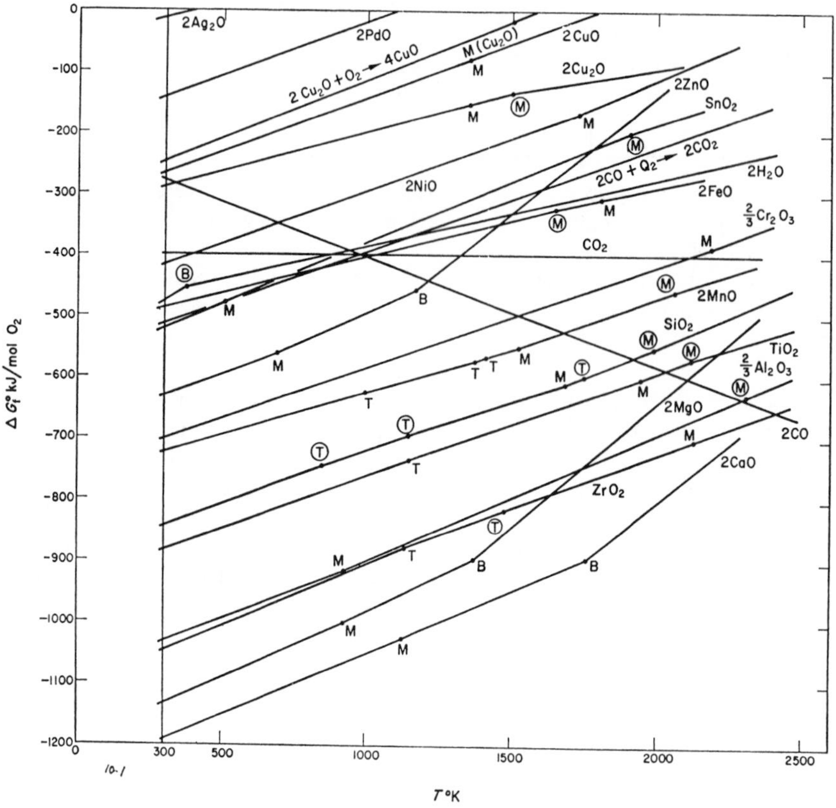

FIG. 10.1 A simplified Ellingham diagram for oxide formation. It shows ΔG_i^0, based on one mole of oxygen, for a number of metal oxides, as a function of temperature. (Based on information made available by the British Iron and Steel Research Association. The letter M, B or T represents the melting, boiling or transition point for the element. A circumscribed O denotes the oxide.)

The first point of interest is that the curves consist of a number of apparently straight lines which change slope abruptly at each phase change. The deviation from actual linearity in most cases is within the experimental errors. The full calculations are not wasted effort, however, because simple extrapolations to high temperatures using low temperature data are not justified. With one extremely important exception, the lines generally have a positive gradient, which means that the oxides become *less stable at elevated temperatures.* Some oxides, such as Ag_2O have $\Delta G^0 = 0$ at the relatively low temperature of 130°C with the result that $K = 1$, and the dissociation oxygen pressure is 1 atm.

The second main point concerns the slope of the lines. In the case of ZnO formation, the line has a positive slope, which steepens at the melting point of zinc (693 K), and again at the boiling point (1180 K). Equation 10.1 relates this slope to ΔS:

$$\left(\frac{d(\Delta G^0)}{dT}\right)_P = -\Delta S^0.$$

The zinc reaction below 693 K,

$$2\,Zn\,(c) + O_2\,(g) \rightarrow 2\,ZnO\,(c),$$

involves the consumption of one mole of gas, and ΔS^0 is clearly negative (it is -200 J/ K at 298 K). At the melting point of zinc, $S^0(Zn)$ increases, and $d(\Delta G)/dT$ becomes more negative as a result. Vaporization of zinc causes a much greater increase of $(-\Delta S)$ and hence of slope. **Steepening is associated with a greater entropy drop on oxide formation.** On the other hand, in the case of Pb/PbO, the oxide boils at 1890 K, and ΔS^0 for the process

$$2\,Pb\,(l) + O_2\,(g) \rightarrow 2\,PbO\,(g)$$

then actually becomes positive because of the two moles of gaseous product. The gradient is negative for a short while until lead boils, when the line sets off uphill once more.

Thirdly, oxidation-reduction processes may be analysed visually, because all reactions are based on one mole of oxygen. At 1000 K the line for Ti/TiO_2 lies well below that for Zn/ZnO. That is, TiO_2 has a larger negative value of ΔG^0, and is the more stable of the two oxides. **The oxide lower on the chart is more stable.** A numerical analysis is simple. For the two oxides, and the reduction, we write:

I	$2\,Zn + O_2 \rightarrow 2\,ZnO,$	$\Delta G^0_{1000} = -489$ kJ,
II	$Ti + O_2 \rightarrow TiO_2,$	$\Delta G^0_{1000} = -761$ kJ.
(II–I)	$2\,ZnO + Ti \rightarrow 2\,Zn + TiO_2,$	$\Delta G^0_{1000} = -272$ kJ.

Thus zinc oxide would clearly be (thermodynamically) unstable in contact with hot titanium, although the rate of such corrosion would be extremely slow, due to the extraordinarily tough oxide layer on titanium metal. Moreover, we see that the use of zinc as a reducing agent for TiO_2 is quite out of the question, *not only at* 1000 K, *but at any practicable temperature,* because the titanium line remains below that of zinc, whatever the temperature.

The only oxide to become consistently more stable as temperature increases is carbon monoxide. On this fact the iron age depended; virtually any metal oxide is reducible by carbon *at a sufficiently high temperature.* Visually, this means that the carbon monoxide line moves underneath those of other oxides, as the temperature increases. Thus manganous oxide (MnO) is stable to hot coke up to 1680 K, at which point the lines for MnO and CO meet, and ΔG^0 for the reaction

$$MnO + C \rightarrow Mn + CO$$

is zero, since it is equal to $\Delta G^0_f(CO) - \Delta G^0_f(MnO)$. Above this temperature ΔG^0 becomes negative; at 1750 K, $\Delta G^0 = -533.5 - (-508.4) = -25.1$ kJ, and the production of liquid manganese metal proceeds.

It then becomes possible, although economically unattractive, to reduce other oxides with carbon at yet higher temperatures. Production of silicon becomes feasible at 1800 K, of titanium at 1860, of magnesium at 1910, and even of aluminium, zirconium and calcium between 2350 and 2450 K. Although both titanium and magnesium have been produced by the 'carbothermic' method, the cost of such high-temperature operations is very high, and alternative processes are now in use. In particular, electrolytic methods become increasingly attractive in this situation. A free energy change of 600 kJ/mol O_2 may be overcome by a mere 1.6 volts of applied external potential, and so wherever cheap electricity is available, such as in Canada or Ghana (Volta River Project) large aluminium plants may spring up. The *equilibrium* potential will be simply $E^0 = -\Delta G^0/nF$, although higher voltages than this are invariably used. For an appreciable current, the cell resistance and electrode polarization must be overcome; the resistive heating is often sufficient to keep the electrolyte molten.

Carbon has two oxides which are of comparable stability, and as a result shows dual behaviour. Formation of monoxide, as we have seen, gives a downward sloping line. The formation of carbon dioxide,

$$C + O_2 \rightarrow CO_2,$$

has $\Delta S^0 = +2.9$ joule/K, and gives an almost horizontal line, which cuts that of the monoxide at 983 K. Below this temperature formation of CO_2 predominates; above it, CO. At 983, they are sharing the honours. The (Boudouard) disproportionation reaction

$$C + CO_2 \rightarrow 2\ CO$$

has $\Delta G^0 = \Delta G_f^0(2\ CO) - \Delta G_f^0(CO_2) = 0$, and $K = 1$. An additional line, representing the oxidation of CO,

$$2\ CO + O_2 \rightarrow 2\ CO_2,$$

is important during low temperature reductions, because CO is the gas phase intermediate. The monoxide runs a shuttle service for oxygen between the solid oxide and the solid-coke reducing agent, but is not itself an overall reaction product.

We are now able to discuss four applications, two briefly, and two in some detail.

10.3.1 Corrosion Prevention

Steam turbines can become more efficient with hotter primary steam. This in turn requires high temperature furnaces, and in nuclear power stations, fuel elements that can withstand higher temperatures. Liquid sodium has been used in nuclear reactors as a primary, closed-loop, cooling medium, but the possibility of its corroding the stainless steel used to contain the uranium oxide fuel elements has caused concern.

Figure 10.2 shows clearly that at about 1000°C sodium oxide can oxidize stainless steel (the dotted line represents liquid sodium above its normal boiling point). A diagram of this kind shows not only that sodium can cause material transport of oxygen from a point of external contamination, but also that metallic zirconium is able to 'hot-trap' the oxygen from the sodium stream.

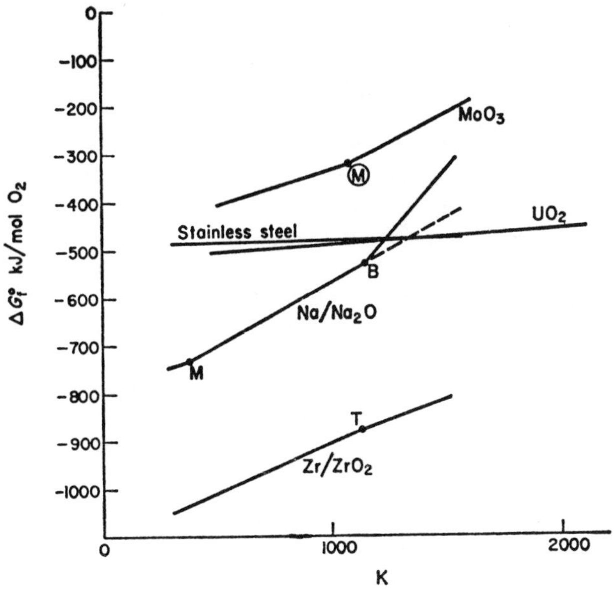

FIG. 10.2 Ellingham free energy diagram showing possibility of corrosion of stainless steel fuel cans by sodium oxide near 1250 K. Liquid sodium coolant may be kept free of oxygen by zirconium metal, which has greater affinity for oxygen.

10.3.2 Electrolysis of Alumina

Paul Louis Toussaint Héroult obtained a patent in April 1886, just seven weeks before the American Charles Hall obtained his, for processes to make aluminium from bauxite. They were independently developed, but almost identical. A molten bath of bauxite, Al_2O_3, in cryolite, $Na_3[AlF_6]$, is electrolysed with carbon electrodes at about 1240 K. Molten aluminium is released at the cathode, and the oxide ions released at the anode oxidize it to a *non-equilibrium mixture* composed mainly of carbon dioxide; the overall reaction is

$$\tfrac{2}{3}\,Al_2O_3 + C \longrightarrow CO_2 + \tfrac{4}{3}\,Al.$$

If this cell were to operate reversibly, the e.m.f. would depend on the free energy change for the process, which is $\Delta G^0(CO_2) - \Delta G^0(\tfrac{2}{3}\,Al_2O_3)$. At 1240 K, this is equal to $-397 - (-841) = +444$ kJ. Four electrons are involved in the discharge of two oxide ions, so $E^0 = 444/4(96.487) = 1.15$ V. This e.m.f. must be exceeded in order to produce aluminium at a measurable rate. In fact, the cells are operated at about 5 volts. Notice that carbon cannot itself reduce Al_2O_3 at the operating temperature, but it does help. On the other hand, the use of a non-reacting anode would require greater electrical energy; the process as a whole would be unnecessarily expensive.

10.3.3 Thermal Reduction of Magnesia

Magnesium is the sixth most abundant metal in the earth's crust, but is exceedingly reactive. It finds wide use as a light alloying material, and is extensively used in the aircraft industry. Although traditionally produced by electrolytic methods, it requires between 8 and 9 kWh of electricity per pound of metal. Thermal methods are now widely used.

During the 1939–45 war, the simple carbon reduction process was operated. The process is interesting, although obsolete:

$$2\,MgO\,(c) + 2\,C\,(c) \rightleftharpoons 2\,CO\,(g) + 2\,Mg\,(g).$$

From the oxide chart, it is seen that a temperature of over 2000 K is required to obtain negative values of ΔG^0. The cost of operating at such temperatures was high, and in addition separation of the gaseous products was awkward. On cooling the mixture, the back reaction occurred, and so shock cooling with hydrogen was resorted to. This gave a finely powdered product, which inflamed on contact with air. Demand for magnesium was so high at that time, however, that even this inelegant process was pressed into service.

Examination of the oxide chart shows that silicon, although not a good reducing agent at high temperatures, is better than carbon below about 1750 K. Having said this, it still does not show great promise in the case of magnesium. However, by making two vital modifications, L. M. Pidgeon [6] was able to show that silicon reduction was an economic possibility; it is now widely used wherever electrical energy for electrolysis is expensive. The first improvement involves the silica product of the main reaction

$$2\,MgO + Si \longrightarrow 2\,Mg + SiO_2. \tag{10.9}$$

Dolomite, $MgCO_3.CaCO_3$, occurs widely; on calcining, it yields the mixed oxide. On using this in place of MgO, the activity of the silica product is much reduced by formation of calcium silicate, $CaSiO_3$. This helps formation of magnesium, which is gaseous at the temperature of commercial operation, 1450 K. The silicate formation reaction,

$$CaO + SiO_2 \longrightarrow CaSiO_3,$$

contributes approximately -92 kJ to ΔG^0 for the overall process, which is

$$CaO + 2\,MgO + Si \longrightarrow CaSiO_3 + 2\,Mg. \tag{10.10}$$

Using the oxide chart, ΔG^0 for eq. 10.9 is found to be 218 kJ, and so for eq. 10.10,

$$\Delta G^0_{1450} = +218 - 92 = \mathbf{+126\ kJ.}$$

The second modification concerns the pressure. Although ΔG^0_{1450} is $+126$ kJ, *magnesium is the only gas*, and can be pumped off if the total pressure is low enough. We can calculate ΔG for the actual process by using the general reaction isotherm,

$$\Delta G = \Delta G^0 + RT \ln Q.$$

The reaction mixture, which is calcined dolomite, ferrosilicon (75% Si) and 2 per cent calcium fluoride as catalyst, is sealed into steel retorts which are evacuated to a pressure of about $(2)10^{-5}$ atm. On heating to 1180°C, the Mg pressure rises to about $(5)10^{-4}$ atm, and because $Q = p^2_{Mg}$ (all other activities being unity),

$$\begin{aligned}\Delta G &= +126\,000 + 2.303(8.314)(1453) \log ((5)10^{-4})^2 \text{ joules}\\ &= +126 - 184 = -58 \text{ kJ.}\end{aligned}$$

Because the magnesium is distilled, the product is extremely pure. A typical analysis is shown in Table 10.2.

TABLE 10.2 *Typical analysis of magnesium obtained by the Pidgeon process*

Impurity	Si	Al	Fe	Cu	Mn	Ni
Content, p.p.m.	40	40	20	10	10	5

This process depended for its development on a clear understanding of free energies and equilibria, and adequately demonstrates the power of such methods.

10.3.4 Titanium and the Kroll Process

Ellingham diagrams are available for many types of reaction, such as the formation of carbides, nitrides, oxides, sulphides, chlorides, and fluorides [7], and the dissociation of sulphates and carbonates [8]. Figure 10.3 shows a simplified Ellingham diagram for chlorides, based on one mole of chlorine, Cl_2. It is impressive not only that chlorination reactions may be studied, but that information from two or more diagrams may be combined together.

Titanium is superbly corrosion-resistant, and has the best strength-to-weight ratio between 200 and 400°C of any commercially available metal. It is expensive to produce in a pure state because methods discussed so far cannot be used. Thus carbon reduction of the oxide leaves some titanium carbide as impurity, and electrolytic processes meet with the difficulty that $TiCl_4$ is non-conducting. In fact, titanium is formed [9] from rutile, TiO_2, by a two-stage process of chlorination to form $TiCl_4$ followed by reduction of this intermediate by a reactive metal. The reactions are

$$TiO_2 + 2\,C + 2\,Cl_2 \rightarrow TiCl_4 + 2\,CO \qquad (10.11)$$

and

$$TiCl_4 + 2\,Mg \rightarrow 2\,MgCl_2 + Ti. \qquad (10.12)$$

Reaction 10.11 involves the oxides of titanium and carbon, and the chloride of titanium; thus we shall make use of both the chloride and the oxide diagrams. The reaction takes place between 1000 and 1250 K. At 1100, ΔG^0 for eq. 10.11 is

$$\begin{aligned}\Delta G^0_{1100} &= 2\Delta G^0(\tfrac{1}{2}\,TiCl_4) + \Delta G^0(2\,CO) - \Delta G^0(TiO_2)\\ &= 2(-314) + (-416) - (-747)\\ &= \mathbf{-297\ kJ.}\end{aligned}$$

Notice that each value of ΔG^0 refers to one mole of O_2 or Cl_2 as appropriate. Notice also that were it not for the reducing effect of the carbon, the reaction would not be possible. (There are several metal oxides, for example those of copper, iron, and zinc, which may be chlorinated by chlorine alone.) The $TiCl_4$ is a volatile, essentially covalent liquid with a boiling point of 136°C, which must be of high purity if a high quality titanium product is to be obtained.

The magnesium reduction process, represented by eq. 10.12, was developed by W. J. Kroll. Figure 10.3 shows that $TiCl_4$ is reducible by aluminium, mag-

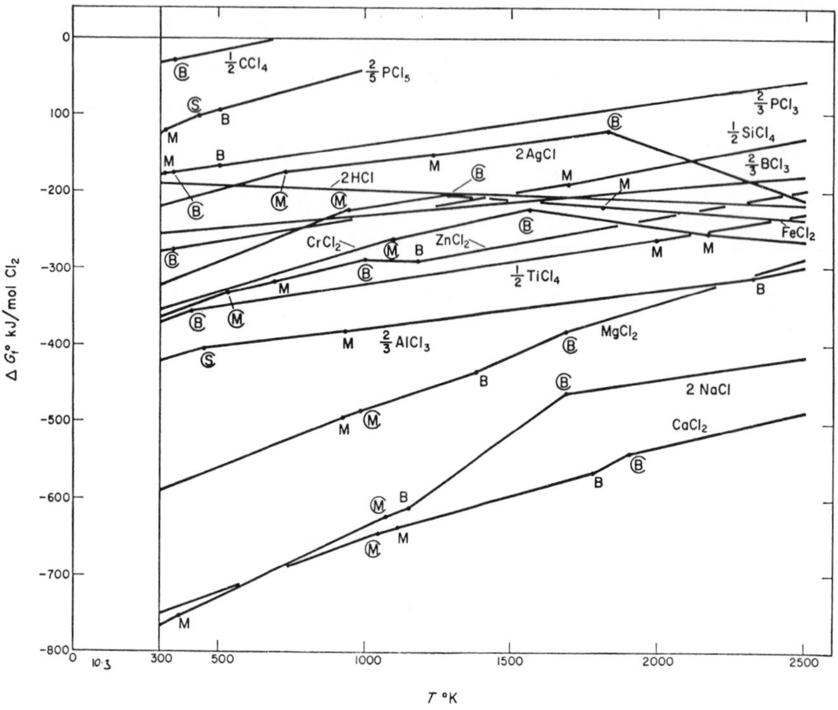

FIG. 10.3 Ellingham diagram for the formation of a selection of chlorides, based on one mole of chlorine. (Glassner, A., 'The thermochemical properties of the oxides, fluorides and chlorides to 2500 K', *Report ANL 5750*, United States Atomic Energy Commission. M, B or T represents the melting, boiling or transition point for the element. A circumscribed C denotes the chloride.)

nesium, sodium, or calcium. The choice is governed not only by the cost of the metal, but also by the ease with which the titanium metal may be separated from the other products. Magnesium was chosen initially. To molten magnesium in a steel pot, kept under an inert gas, titanium tetrachloride is slowly added. The reaction is strongly exothermic ($\Delta H = -484$ kJ) and the temperature is kept below 1150 K. From the diagrams, $\Delta G^0 = -318$ kJ. Notice that $MgCl_2$ is molten at this temperature; it is run off from time to time. The titanium 'sponge' is porous and contains $MgCl_2$ and unreacted magnesium which must be removed by leaching with dilute acid.

Alternatively, a similar procedure is followed by I.C.I. who use liquid sodium metal. A disadvantage is that there is a very narrow range of temperature between the melting point of sodium chloride and the boiling point of sodium. On the other hand liquid sodium is more easily piped, and the leaching may be done by water. The free energy change is substantially more negative than that using magnesium, although this confers no advantage on the process.

Summary This chapter has presented a set of variations on the theme of free energy changes. We have seen how accurate values of ΔG^0 may be derived at any temperature, how this data may be summarized either as *free-energy functions* or in terms of *Ellingham diagrams*, and how the data may be applied in a few instances. In virtually all cases, we have seen that activity can simplify the calculations of equilibrium constants and that allowance can always be made for non-ideal behaviour, assuming that activity coefficient data are available. Complete thermodynamic data have been published for relatively few compounds, however, and there are for example many common organic compounds for which only an enthalpy of formation has been determined. As more complete information is circulated, the number of applications of chemical thermodynamics will continue to increase.

REFERENCES

[1] Lewis, G. N. and Randall, M. *Thermodynamics*, 2nd edition, Ch. 15 and Appendix 7, McGraw-Hill, New York (1961).
[2] Shell Development Company, Emeryville, California.
[3] Bakshi, Yu. M., Gel'bshtein, A. I. and Temkin, M. I. *Doklady Akademii Nauk SSSR* **132,** 157 (1960).
[4] Ellingham, H. J. T. *J. Soc. chem. Ind.* **63,** 125 (1944).
[5] Ives, D. J. G. *Principles of the Extraction of metals,* Monograph **3,** Royal Institute of Chemistry, (1960).
[6] Pidgeon, L. M. *Trans. Canad. Min. Inst.* **49,** 621 (1946), and *Disc. Faraday Soc.* **4,** 197 (1948).
[7] Smithells, C. J. *Metals Reference Book*, Vol. 2, Butterworths, London (1955).
[8] Hopkins, D. W. *Physical Chemistry and Metal Extraction,* 192, Garnett Miller, London (1954).
[9] Gray, J. J. and Carter A. *Chemistry and Metallurgy of Titanium Production,* Royal Institute of Chemistry, Lectures, Monographs, and Reports, **1** (1958).

PROBLEMS

10.1 Assuming that ΔH and ΔS are temperature-invariant, make an approximate calculation of K_a for the reaction: $CO + 2H_2 \rightarrow CH_3OH$ at 350°C, using tabulated data.

10.2 An accurate expression for ΔG^0 for the methanol synthesis (see previous question) is

$$\Delta G^0 = -73\,350 + 76.11T \ln T - 0.059\,0T^2 - 252.7T.$$

(a) Find K_a at 350°C (623.2 K).

(b) First, assuming that $K_a = K_p$ ($K_\gamma = 1$) calculate the partial pressure of methanol in an equilibrium mixture held at 250 atm. The reactants are in the stoichiometric proportion of 1 : 2, $CO : H_2$. What proportion of CO is converted?

(c) Using generalized fugacity charts, K_γ is found to be 0.41 under the actual operating conditions. Calculate K_p in this case, and thence the equilibrium pressures of CO, H_2, and CH_3OH. What is the CO conversion?

10.3 In some nuclear reactors, UO_2 fuel elements in stainless steel cans are subjected to high rates of 'burning'; that is, substantial quantities of fission products are formed. As a result, appreciable amounts of O_2 are liberated (from UO_2) which will then react either with the fission products (zirconium, molybdenum, cerium, etc.) *or with the stainless steel of the can.* Decide, with the help of Fig. 10.2 whether zirconium and molybdenum are oxidized in preference to the steel ($T = 750°C$). (In fact sufficient reactive elements are formed to take up the oxygen.)

10.4 Using Fig. 10.2, decide at what temperature, if any, UO_2 fuel oxidizes the stainless steel can. (See previous example.)

10.5 Magnesium has been produced by the following reaction:

$$\tfrac{4}{3}\,\mathrm{Al\,(l)} + 2\,\mathrm{MgO} \longrightarrow 2\,\mathrm{Mg\,(g)} + \tfrac{2}{3}\,\mathrm{Al_2O_3}.$$

Process conditions commonly used were $T = 1200°C$, and $P = 0.02$ atm. First calculate ΔG^0 for the process, from the oxide Ellingham diagram, then use the reaction isotherm to calculate ΔG under operating conditions.

10.6 Calcium is produced by the electrolysis of molten calcium chloride. What is the minimum (equilibrium) voltage needed for electrolysis at the melting point of the chloride? (Use Fig. 10.3.)

10.7 Manganese is produced by carbon reduction of pyrolusite, MnO_2, which occurs in two stages; $MnO_2 \longrightarrow MnO \longrightarrow Mn$. Determine the lowest practicable temperature for the second stage of the process. Is the product easily separable from the reaction mixture?

10.8 Pure silicon may be prepared by the reaction $SiCl_4 + 2\,Zn \longrightarrow 2\,ZnCl_2 + Si$. Find from the chloride diagram the temperature at which separation of products is convenient. What is K_a approximately at your chosen temperature?

10.9 Magnesium is often produced by electrolysis of a molten mixed chloride electrolyte containing $MgCl_2$, at 750°C. Molten magnesium is released around the cast steel cathodes. Use an Ellingham diagram to determine the equilibrium potential necessary for decomposition.

In fact, 7.54 kWh are used to produce one pound (454 g) of metal. What is the actual potential used, and why is such an excess voltage necessary?

Suggested Further Reading

Two introductory texts on about the same level as this book, are:

MAHAN, B. H., *Elementary Chemical Thermodynamics*, Benjamin, New York (1963).

BAUMAN, R. P., *An Introduction to Equilibrium Thermodynamics*, Prentice-Hall, Englewood Cliffs, N. J. (1966).

An excellent literature survey is given as a 'resource paper':

NASH, L. K. *J. chem. Educ.* **42,** 64 (1965).

The following more advanced textbooks are recommended for their clarity:

DENBIGH, K., *The Principles of Chemical Equilibrium,* Cambridge University Press (1955).

CALDIN, E. F., *An Introduction to Chemical Thermodynamics,* Oxford University Press (1958).

LEWIS, G. N. and RANDALL, M., revised by PITZER, K. S. and BREWER, L., *Thermodynamics,* 2nd edn., McGraw-Hill, New York (1961).

WALL, F. T. *Chemical Thermodynamics,* 2nd edn., Freeman, San Francisco (1965).

GUGGENHEIM, E. A., *Elements of Chemical Thermodynamics,* Monographs for Teachers, **12,** Royal Institute of Chemistry (1968).

Other more specific references apply to some sections of this book:

Energy, Chapter 1. UBBELOHDE, A. R. *Man and Energy,* George Braziller, New York, (1955).

Propellants, Chapter 2. GLASSMANN, I. *Am. Sci.* **53,** 308 (1965).

Fuel cells, Chapter 9. AUSTIN, L. G. *Scientific American* **201,** 72 (October 1959).

Ethanol, Chapter 10. CARLE, T. C. and STEWART, D. M. *Chemy Ind.* 831 (1962).

Magnesium, Chapter 10. BALL, C. J. P. *J. Inst. Metals.* **84,** 399 (1956).

Appendix I
The Twenty Most Useful Equations

If thermochemical data are the flesh and blood of chemical thermodynamics, then it is true to say that the bone structure is made up of a small number of defining and operating equations. While to make such a selection is necessarily arbitrary, it is felt that to present it is the best way of demonstrating the underlying structure of the subject.

These equations have already been emphasized in the text.

$\Delta U = q_{\text{to}} + w_{\text{on}}$; defining equation for U. (2.1)

$H = U + PV$; defining equation for H. (2.5)

$\Delta H = \Delta U + P\Delta V$ (P constant). (2.7)

$\Delta H = \Delta U + \Delta nRT$ (P, T constant); n refers to number of moles of gas. (2.9)

$(\mathrm{d}(\Delta H)/\mathrm{d}T)_P = \Delta C_P$; Kirchhoff's equation. (3.4)

$\mathrm{d}S = \mathrm{d}q/T$ (reversible change), (5.1)
$\mathrm{d}S > \mathrm{d}q/T$ (irreversible change); formal definitions of entropy.

$\Delta S = nR \ln V_2/V_1$ (T constant, ideal gas). (5.3)

$$\Delta S = \int_{T_1}^{T_2} \frac{C_P}{T} \mathrm{d}T \text{ (}P\text{ constant).} \qquad (5.5)$$

$$S = \frac{R}{N} \ln \omega = k \ln \omega. \qquad (5.10)$$

$G = H - TS$; defining equation for G. (6.1)

$\Delta G = \Delta H - T\Delta S$ (T constant). (6.3)

$\Delta G = \Delta G^0 + RT \ln Q$; general reaction isotherm. (7.13)

$\Delta G^0 = -RT \ln K_a$; equilibrium isotherm. Standard free energy change is directly linked to equilibrium constant in activities. (7.14)

$$\ln K_a = -\frac{\Delta H^0}{RT} + \frac{\Delta S^0}{R}\text{; reaction isochore.} \qquad (8.2)$$

$$\frac{\mathrm{d}P}{\mathrm{d}T} = \frac{\Delta H}{T\Delta V}\text{; Clapeyron equation.} \qquad (8.8)$$

$\Delta G^0 = -nFE^0$. (9.2)

$$E = E^0 + \frac{RT}{nF} \ln \frac{a_{\text{OXIDIZED STATES}}}{a_{\text{REDUCED STATES}}}. \qquad (9.6)$$

$$\Delta S^0 = nF\left(\frac{\mathrm{d}E^0}{\mathrm{d}T}\right)_P. \tag{9.11}$$

$$\Delta G = \Delta H + T\left(\frac{\mathrm{d}(\Delta G)}{\mathrm{d}T}\right)_P; \text{ the Gibbs-Helmholtz equation.} \tag{10.2}$$

$dG = -SdT + VdP$

Appendix II. Fundamental Constants and Conversion Factors

Volume of 1 mole ideal gas at s.t.p.	$V = 22.413$ l/mol
Universal gas constant	$R = 1.987\,2$ cal/mol K
	$= 0.082\,05$ l atm/mol K
	$= 8.314$ joule/mol K
Faraday constant	$F = 96\,487$ coulomb/equiv
	$= 23\,061$ cal/volt equiv
Ice point	$0°\text{C} = 273.15$ K
Avogadro's number	$N = (6.023)10^{23}$ particles/mol

1 joule $= 10^7$ erg $= 1$ volt coulomb
$= 0.009\,87$ l atm.

1 cal $= 0.041\,29$ l atm.
$= 4.184$ joule (exactly) (defined)

1 mole $= N$ particles

$\ln x = 2.302\,6 \log x$

$2.3026\, RT/F = 0.059\,16$ volt at 298.15 K

Answers to Problems

Chapter 1

1.1 0 K = −273.17°C.
1.2 73.2 J.
1.3 1610 g fuel oil.

Chapter 2

2.1 (a) $\Delta H = -56.9$ kJ.
(b) $\Delta H_f^0 = 15.48$ kJ/mol.
2.2 7.32 kJ/mol.
2.3 −37.7 kJ/mol.
2.4 $\Delta H_{isom} = 0, -5.44, -3.14, -8.16, -14.43$ kJ/mol, respectively.
2.5 17.66 kJ/g.
2.6 −148.7 kJ/mol.
2.7 −158.3 kJ/mol.
2.8 502 J/mol $CHCl_3$.
2.9 Heat of combustion, kJ/g = 50.33 (C_3H_8), 49.54 (C_4H_{10}).
2.10 −718.4 kJ/mol.

Chapter 3

3.1 $w = 0$, $q_V = \Delta U = -4289$ kJ/mol, $\Delta H = -4285$ kJ/mol.
3.2 −92.93 kJ/mol.
3.3 $\Delta C_P = 33.1$ J/mol K, $\Delta H^0_{1500°C} = 425.1$ kJ/mol.
3.4 417.6 kJ/mol
3.5 (a) 17.87 kJ/mol (b) $(2.96)10^3$ kg/s.
3.6 58.97 kJ/mol.
3.7 7200 K.
3.8 Cubic equation is

$$-\Delta H^0 = 757\,040$$
$$= 112.5T + (21.0)10^{-3}T^2 - (20.14)10^{-7}T^3 - 35\,355;$$
$$T = 4740 \text{ K}.$$

3.9 −125.1 kJ/mol.
3.10 50.2 kJ.

Chapter 4

4.2 224.6 J/mol K.

Chapter 5

5.1 −ve, +ve, +ve, zero, −ve.
5.2 $(2.68)10^4$ J/K.

5.3 $\Delta H = -11.40$ kJ, $\Delta S = -41.76$ J/K.
5.4 21.0 J/mol K.
5.5 37.2 J/mol K.
5.6 22.1 (WF_6), 30.2 (OsF_6), 26.0 (IrF_6) J/mol K.

CHAPTER 6

6.1 $\Delta S = 26.8$ J/K, $\Delta G = -8.74$ kJ.
6.2 $SnCl_4$ 376(387), cyclohexane 342(354), acetone 331(329), ethane 181.4(184.9) K. (Actual values in parentheses.)
6.3 $\Delta G_{623} = -43.93$ kJ/mol, $T = 1643$ K.
6.4 $\Delta G_{623} = -206.1$ kJ/mol. Reaction would occur even at reduced activity of Na_2O in glass.
6.5 $\Delta G^0_{900} = -19.58$, $\Delta G^0_{1150} = 2.176$ kJ/mol. PdO loses oxygen above 1000 K.
6.6 -99.87 kJ/mol.
6.7 (a) 413 J/mol K. (b) Back reaction occurs at lower temperatures. (c) No net bonds made or broken; ΔH^0_f(Si (g)) is 444 kJ/mol, so ΔH is reasonable.
6.8 Assuming ΔH constant, $\Delta S = 193.9$ J/mol K. Positive ΔS indicates high-temperature occurrence.

CHAPTER 7

7.1 (a) 241.3 kJ/mol SiH_4. (b) -25.19 kJ/mol SiH_4.
7.2 (a) $\Delta G_{\mathrm{I}} = -250.3$, $\Delta G_{\mathrm{II}} = -262.0$ kJ.
(b) -11.7 kJ/mol SiO_2.
7.3 (a) $\Delta G^0 = -110.6$ kJ, $K = (1.81)10^6$.
(b) $\Delta G^0 = -573.5$ kJ, $K = (3.4)10^{23}$.
(c) $\Delta G^0 = -14.23$ kJ, $K = 2.62$.
7.4 $K = 4.865$, $\Delta G^0 = -14.06$ kJ/mol CO_2.
7.5 $K = 0.1563$, fraction dissociated $= 0.194$.
7.6 $\Delta G = -349.4$ kJ/mol from reaction isotherm.
7.7 $\Delta G_{\text{vaporization}} = +9.43$ kJ/mol H_2O.
Specified process has $\Delta G = 10.90$ kJ/mol, giving $p = (1.106)10^{-2}$ atm $= 8.41$ mm Hg.
7.8 $(1.45)10^{-14}$ atm.
7.9 26.40 kJ/mol. ΔG^0_{298} is promisingly negative, but rates are very slow.
7.10 $\Delta G = -1.665$ kJ, $K = 1.36$.

CHAPTER 8

8.1 207.9 kJ/mol.
8.2 $\ln p_{\text{mm}} = -42\,740/RT + 20.409$, 88.2°C.
8.3 $\Delta H_{\text{vap}} = 68.37$, $\Delta H_{\text{sub}} = 137.9$ kJ/mol, $T_M = 392$ K.
8.4 113.0 kJ/mol.
8.5 For $CO_2 + C \rightarrow 2\,CO$, $\Delta G = -32.77$, $\Delta H = 178.0$ kJ/mol at 1173 K.
For $2\,CO_2 \rightarrow 2\,CO + O_2$, $\Delta G = 357.1$, $\Delta H = 568.7$ kJ/mol at 1173 K, and $\Delta S = 180.3$ J/mol K.
8.6 $\Delta G^0_{524} = -4.301$, $\Delta G^0_{580} = -15.38$, $\Delta H^0 = 97.28$ kJ/mol.
8.7 -40.50 kJ/mol.

8.8 $\Delta H^0_{f,2000} = -247.5$ kJ/mol, DH^0(B—N) = 628 kJ/mol.
8.9 ΔH^0_{298} for neutralization = −56.5 kJ/mol.
Using ΔH^0_f values for $H^+ + OH^- \rightarrow H_2O$ (l),
$\Delta H^0_f(OH^-) = -229.3$ kJ/mol.
8.10 119.23°C.
8.11 0.219 atm.

CHAPTER 9

9.1 $(3.1)10^{-27}$.
9.2 (a) $Fe + 2\,OH^- \rightarrow FeO + H_2O + 2e$,
$Ni_2O_3 + H_2O + 2e \rightarrow 2\,NiO + 2\,OH^-$.
(b) ΔG(cell) = −245.1 kJ/mol Fe,
$\Delta G^0_f(Ni_2O_3) = -431.8$ kJ/mol.
9.3 0.725 V.
9.4 (c) $E^0 = +0.0528$ V.
9.5 $E = +0.020\,7$ V.
9.6 95 900 coulomb/equivalent.
9.7 0.071 3 V.
9.8 −0.020 3 V.
9.9 −65.5 J/mol K, cell value.
9.10 $\Delta G = -20.05$, $\Delta H = -25.04$ kJ/mol Ag, $\Delta S = -18.2$ J/mol K. Solid components not pure crystals, activities non-zero.
9.11 (a) 30.9 J/mol K. (b) 8.12 kJ/mol.
9.12 $\Delta G = -98.28$ kJ/mol Hg, $\Delta S = -3.9$ J/mol K.

CHAPTER 10

10.1 $\Delta G^0_{623.15} = 48.24$ kJ/mol CO, $K = (9.02)10^{-5}$.
10.2 (a) $(4.875)10^{-5}$.
(b) $p_{CH_3OH} = 54.2$ atm, % conversion CO = 45.4.
(c) $p_{CH_3OH} = 82.6$, $p_{CO} = 55.8$, $p_{H_2} = 111.6$ atm,
% conversion CO = 59.7.
10.3 Steel oxidizes after Zr, before Mo.
10.4 >1400 K.
10.5 $\Delta G^0 = 54.4$, $\Delta G = -41.4$ kJ/mol O_2.
10.6 3.34 V.
10.7 1684 K is lowest temperature; Mn is liquid, MnO solid. Separation convenient 1684–1900 K.
10.8 Process performed at 1000°C (1273 K), at which $\Delta G^0 = -64.0$ kJ/mol, $K = 423$.
10.9 $E^0 = 2.48$ V. 7.5 V actually used.

Index